MODERN CONSTELLATIONS

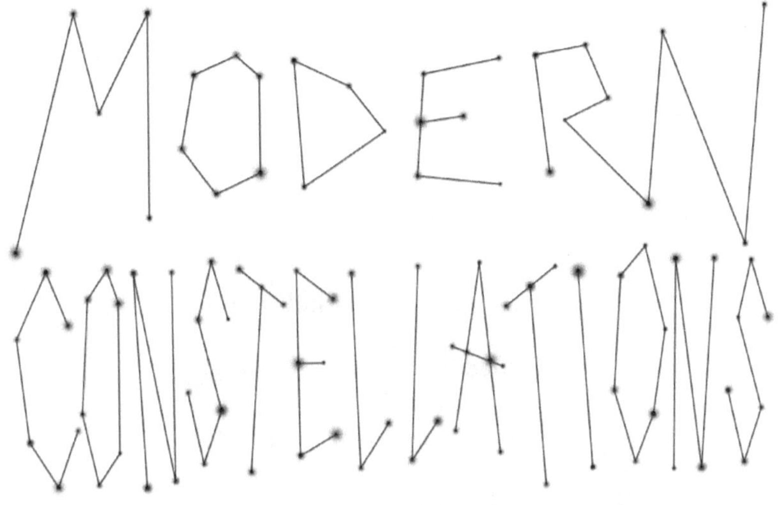

MODERN CONSTELLATIONS

KENDALL NICHOLS

atmosphere press

Table of Contents

Lesson 5: The stars aren't that far or wrong

Bonus poems

Introduction

The book is not a conventional poetry and or book of short stories. The book features traditional poetry styles, how-to poems, converted slam poetry pieces, short stories, and everything in between. Like my mind, this book covers a lot of topics and does this in a variety of ways.

Who am I? My writing speaks for itself. If you never met me, then my writing will be the next best thing. The best way to describe me is through my clothing and style. My wardrobe, like my poetry, does not have one set style it follows.

Dedications

To my mom, cuz duh.

To Mr. Sarudi, my seventh grade English teacher who said my writing was unique.

To Mrs. Sparagis, my eighth grade English teacher who made me love poetry.

To the people whom I met and helped shape my life.

To the people who said I wasn't good enough.

To the people who made me experience the highs and lows of life.

To the people who know that life isn't what other people think it is.

To anyone who feels the pressure and still smiles when life isn't kind.

To anyone who thinks they're not good enough, you are more than enough.

~This is dedicated to everyone~

LESSON 1:

The roads traveled aren't the same

My melanin always reminds me of leather.

It's durable and long lasting, but apparently

it's bad for the environment.

To the Reflection in the Mirror

She says she's crazy,
okay.
She says she's emotional, close to a wreck
the *Titanic* can't compare.
She says she's a handful
and I'm a hands-on person.
I'm waiting for her to say something that drives me away
like driving into a sunset,
tonight it's beautiful
and so is she.
Was she,
as she compares past pictures of herself with agony.
Tenses never agreed with me
I always tensed up around her, cuz
what do you do when you see an angel in person?
I smile, I smiled
I'm smiling thinking of her,
please make this stop.
I think I'm lost.
I got lost in a maze of wondering how she exists
of how she still stands when the world wants to break her.
She keeps talking,
giving numerous reasons for how and why I could break her
I couldn't even break a sweat, let alone a heart
her heart.
You think if I asked her to jump off a cliff with me, she would
	say yes?
Either we sink and someone else saves our lives
or we swim to shore and laugh about the fall.
She says she's not the one for me
I laughed,
I don't remember asking her to be.
How do you tell someone that the risk is worth more than the
	regret?
You keep staring at me expecting me to do something
I can't do anything if her mind is like her heart
closed
to me.

Maybe I'll open them both,
but for now
I'll smile and say okay.

Multiples Haikus and Tankas

I hate the cold, and
what it represents for us;
Yet I'm here for you.

Why did you leave us?
You never really liked snow.
Were we not enough?
The family was broken.
Are you okay (you don't text)?

People wish for spring.
Spring showers don't bring flowers.
Spring brings excuses.

Summer came, not you.
Family, you said you'd return.
Stopped by for minutes,
no longer than time allowed.
Now I wish for cold, alone.

———————————

What is depression?
Boy asks his tired mother
She can not respond

Words can not describe
The insurmountable pain,
Days of not wanting,
Days of being lost in daze,
Where the sun shines, not for her.

What is sacrifice?
Boy asks absentee father
Father can't respond

"For the greater good,"
As the mother gives response,
"It requires demand,
staying when you don't want to,
It is being there when asked."

Lighthearted ways to see some devices

Cliché
Realize, real eyes, real lies
When you realize with your real eyes, you can truly see the real lies and not be blinded enough to see where all the real lines are

Corny
When I kissed her, I knew she wasn't the one.

Copyright infringement
"Seeing is deceiving, dreaming is believing, it's okay to not be okay."

It's gonna get better, things are going to turn around.
Plain *bullshit* that we tell ourselves because lot of us can't handle the truth of reality.

Personification
The sun is almost as beautiful as my bed
I bought a wooden whistle. But it wouldn't whistle.
A bad *pun* that should never be repeated.

The sky was painted with a thousand different blues as the clouds which remind me of cotton candy connected to touch the tips of mountain tops in a great plain where the grass danced in the wind.
Imagery that should be reminding people that life is short, and it may be hard but there are moments everyday which we can look to remember that life is beautiful and not take for granted.

If the sun has a million degrees, then why did it still go to college?
An awful **joke** which can represent how the sun isn't infinite and how it will one day consume our entire existence, nothing scary just extremely ordinary.

Go ahead and give them a show,
talk about anything maybe *Wade v Roe*

cuz that's a devastating blow
or how you never felt you could be this low,
talk about what you don't know.

Consonance: a way to express topics while using words that
stress certain syllables so that anyone reading or listening will
remain constant.

This country is a black hole, it takes any good idea and destroys
it
Metaphor, believing that the gravity of importance
needs to be understood between two different ideas
before we're all consumed.

It's not as bad as a crooked smile.
A **simile** stating that crooked smiles are bad
when in reality there's nothing wrong except the preconceived
perception
perpetuates stigmas that stay in the system.

Alliteration: alluding to awful awestruck businesses by bribing
broken confused conflicted demons downright demoralizing
every efficient future firsthand human who had hoped for
heaven on earth.
Translation: big bad businesses will make the future hell and
not heaven.

Spoken word, Yelling! Angry! Waving my hands a LOT!
Specific point of view on THINGS! Cynthia! Cyn-thi-a! Jesus
died for our sin-thi-as! While combining every single English
device in a giant dumpster fire to create a topic that matters to
the poet and therefore should matter to you.

Something real

Roses are red,
Violets are blue,
that poem ain't meant to be true.
Bodies are dead
Violence is blue
People are afraid, now the cops watch you.
Cliché corny catch phrases catching hands
as fast from featherless fallen angels aiding any
robust rebels ready to
spit solid sources of strategic
analytics against troops of anti-truth.
I just wanted to catch your attention
like CPS taking away teachers pensions
How about we talk about something other than somber streets
    where
once blessed beings were butchered by bloody boulevards.
I just wanted to talk about some other issues and not the newest
    Jordan shoes
cuz I don't want you to blow any more tissues;
Blowing tissues over the last brown body dropped on the block
like a dime dropped by a weed dealer.
Can we talk about that really puts you in the feels,
cuz look here's my main deal:
I have a confession,
not the one where you tell the father in little dark chamber
rooms
because he's done much worse to little boys and it's not going to
    get written on his tomb.
My confession is not you not believing in Jesus
because God don't believe in you and that's only half true.
My confession isn't that guns kill people,
it would be people kill people and
guns are just one way to do it.
I'm tired of us all hiding behind fake fallacies
protecting ourselves like the phony pharisees
behind the Hebrews when they get bruised,
like look it's a double lose/lose and now you look worse than

Blue's Clues.
Can we talk about something real,
something that's bigger than your Netflix 'n' chill?
Now look, people been thinking they some halfway crooks
and I'm not having anyone trying to pretend that if dope comes,
    they can cook.
Let's not talk about the same issues,
let's talk about what isn't truth and how we can make it right,
Because this world is a little tight.
Too tight in that most people can't look past race
No haste all waste worrying about these waists
instead of contemplating on how to save face
So can we as a community of people talk about something real,
    something that's not ideal?

Pau's story

"You ever read the poem 'Road Not Taken' by Robert Frost? I think he got it all wrong. Instead of telling people to take those roads and their lives could end up differently, he should've meant something else. It would've been better if he had spoken about how life is made up of different roads and in the end we all end up in the same place, dead. Or, if he wanted to go lighter, he could have said that we shouldn't worry what road we take, because all of our paths are different. Or if he wanted people to change, he could say 'keep the destination and don't worry about the roads not taken, just get there.'"

"You're into poetry. You could've fooled me," I said. I was a sixteen-year-old boy with braces and glasses. I had a short '80s style afro that was unpicked and nappy, though an average person wouldn't be able to tell because I always wore a Chicago White Sox hat.

"I'm not. I'm just trying to pass time."

"Why?"

"Dude, we're like on an airplane, and I can't go to sleep."

"No, not why are you talking to me. Why are you not into poetry?"

She shrugged, "I don't know. Wasn't into that Shakespeare guy in high school and all these forms and rules you have to follow. I don't know if you could tell, but I'm not a fan of following traditions."

"I don't know you very well besides the past forty-five minutes, but I think you would actually like poetry. I like alliteration by the way. What I do know about you is that you're bold, brilliant, and beautiful."

Pau blushed, and silence ensued for the next fifteen seconds.

Who was Pau?

Pau was beautiful. She was the type of girl who guys would go out of their way to make sure she knows how beautiful she was. Pau was strong; she took the world on with one arm and did her makeup with the other. The same arm that took on the world wore a beautiful silver bracelet with roses as clasps. Her life wasn't a walk in the park, it was more like walking through a bed of rose thorn bushes: pretty from afar, but very painful to walk through. Everyone walks with some sort of pain, but Pau somehow came out smiling after every reason not to. Her essence and being was truly inspirational. She hated that word, inspirational.

Fifteen seconds passed and Pau stopped blushing. Her response was worth the wait. "You forgot one. While you were trying to be cute or whatever with that alliteration, you forgot one. Brilliant, bold, beautiful, and badass bitch."

My mouth hung there wide open. I was in disbelief that someone was this full of themselves. A wrong assumption I would make about Pau.

Pau continued, "I didn't say I didn't like poetry, I just said I didn't like to be forced to follow the rules."

16-year-old me responded, "Well that's my fault. I guess I wasn't listening correctly."

"You were listening, you just weren't hearing."

"Excuse me?"

"You seem lost." She then went on to eat a handful of peanuts from the hospitality snacks served from the flight attendant.

"What do you mean I wasn't listening?" My face was something between confusion and bewilderment.

"People listen but they don't hear."

"Saying it again won't help, oh wise buddha," I said.

"See, you were just trying to have a conversation. You took everything at face value. You assumed that I didn't like poetry, when in reality I didn't like the rules. You probably think I'm full of myself and cocky, but I'm not. I'm confident; people who don't think highly of themselves see that as someone being full of themself."

Pau went on for many more minutes describing the difference between listening and hearing and the difference between confidence and cockiness. I was in awe as she told me *listen* and *silent* are made from the same letters. Pau went on to describe that she had to have this talk with her younger brother and that he needs to go after girls who can hear and listen. It sounded like she was describing herself as the type of girl her brother needs to go after. Pau reaffirmed that the girl for her brother wouldn't have to be fluent in Spanish or need to be a wife, mother, and older sister in one. Or go grocery shopping for the house or even be an honor roll student like her.

"So what's he like?"

"My brother or his father?"

With a shrug I asked, "Why not both? I mean, we have time."

Pau emptied the last of her peanuts into her hand and threw them at her mouth with little accuracy. Her black hair flowed down to the middle of her back and was in full shimmer as the sun stared at beauty. The flight was from Chicago to Los Angeles, and the natural sunlight above thirty thousand feet was perfect.

"My brother is only eight, but he has a lot going for him. The girls will be on him in a couple of years. He's sweet and adorable. And he only curses in English so his dad doesn't understand. It's really cute and funny. His dad will go: '*que dijiste?* and then I always say '*no se, es su hijo.*' I laugh. I don't have a lot of opportunities to do that."

"My brother's dad is different. He's a monster. You probably didn't ask, so I'm sorry for this. I would show you the bruises he hands out, but we're on a plane."

"I mean it's not like they can kick us out," I said jokingly.

Pau craned her head and surveyed the aisle. She reached for her shirt and slowly started to lift it up. My eyes widened, and my hand reached hers and pulled her shirt down before she could resist. Our eyes met and hers said, "What?" while mine said, "Chill; I don't really want to see your bruises, any other moment maybe, but not on a plane and not next to my sleeping ten-year-old brother."

I let go of Pau's hand. She straightened her shirt and continued, "My brother's dad isn't anywhere near nice to me. He's not my dad. My mom had me before she met him. They didn't fall in love; she was just tired of barely getting by and having a child. She was too headstrong to ask for my family's help, and it didn't help when they practically kicked her out and disowned her when she had me at seventeen. She met him, had a kid, got cancer, and died. And that's all she wrote."

Pau paused to let me gather in the life advice she gave out before she continued talking, "I could tell you that the piece of shit who is my brother's father always beat our mom. Even when she was pregnant, that didn't stop him. When she died, we all thought he changed. His siblings didn't even try to stop him anymore when he started beating me. When I ran to their houses for help, they would call him and then I would end right back at that house. That didn't break me. It *can't* break me. I refuse to let him or anything break me or stop me from reaching my goals."

How did she end up sitting next to me while I was going through puberty with voice cracks and how old was she?

A random person does not just decide to talk to a sixteen-year-old boy in the tight confines of an aerial vehicle. Before sitting next to me and forcing me to sit in the middle row while my brother had the window, Pau was in the middle of two overweight men. Her eyes locked with mine and pleaded if she could take the empty seat in our row. I moved over right away, and Pau politely told the two gentlemen to suck in their dangling skin so she could get through.

When she sat down, I told her my name and age, and my younger brother laughed. She responded and said she was turning seventeen soon. I naturally asked when because of my invasive personality. To my surprise it was the same day as my sister's. To celebrate we ordered food and drinks.

In between her impressively sized small bites I asked Pau why she was traveling without her brother.

The moments gathered and Pau responded, "Not this time. I literally just got my GED results in the mail. I passed. Oh, I forgot to mention that I dropped out of school in like October or something. No biggie, I've just been working and working out since. My stepdad was pissed that I didn't go to school anymore, but I guess that's understandable. In the one week since I dropped out, we got into so many fights. I couldn't stand living in that house anymore. So, I moved into a place with a friend of my friend. And now I'm done living in The Region, and I'm ready for the LA switch."

Pau was going on a tangent, so I had to step in. In a normal tone I asked, "So what about your brother?"

Pau responded with no hesitation, "I was getting to that point then you interrupted. Rude much." She blew a loud and long deep sigh. "Just because someone is smiling, doesn't mean that they're happy. I'm full of energy right now and talking to you and smiling, but that doesn't mean I'm genuinely happy. I have no obligations anywhere and I would love to take my brother with me, but I just can't. I don't even have a place to crash. I mean I know some people who will let me crash but nothing permanent. And I have no jobs lined up. So, my brother can't be here right now. If my brother was here, then it would be damn near impossible to do anything."

Once again, my mouth hung ajar like a door at an open house. I tried to hide my shock, but every other fact made my eyes widen or made my mouth even more open. "What the—"

"Yeah, I know," Pau said as she interrupted me, "just don't judge."

"You don't know me but trust me I am not judging you. I was going to say: 'what the world, that's awesome.' But you kind of cut me off."

"No judgment like at all. Even if—"

This time I interrupted her, "Even if you're trying to blow the world up. You probably have a good reason, so I can't complain. The point attempting to be made is that it doesn't matter. Your life isn't perfect but, in the end, it's about reaching our goals, not how we get there. I'm positive that your brother will understand why you left."

Pau went back to her small bites but not before I said one last statement that was on my mind, "The poem by Frost, I think that's what the poem should be about. Reaching goals and not the way we achieved them."

So this girl is cool and all, but her real name can't be Pau and pronounced like *paw*?

A couple things to note about Pau: she took life one breath at a time, which has its pros and cons. Pau's real name was Paula and she only shortened it because her mother always called her by that name. As we talked, I found out more about Pau, like her middle name is after her grandmother. Her great-grandmother was the first person to do something "important" and honestly, I can't remember much about it. She never specifically told me her name, she just told me that fact about her great-grandmother, and I never really cared to find out. Her last name never really came up. It's sort of weird to just ask someone their last name.

If anyone wanted to understand Pau, they should know that she wanted to start a new life full of mystery and intrigue. In her new life, if you wanted to know Pau then one would have to figure out everything that she didn't tell them. I love Pau's views on life, but this point of being a complete mystery was excessive and I couldn't disagree more. As the plane began to prepare for landing it made me pay attention to the smaller details. I will never forget the sound of optimism mixed with fear in her voice as we spoke.

"This is where I call bullshit."

"Bullshit on what?"

"On the idea that you have to keep your entire life a mystery to attract those who are 'truly' worthy of wanting to know you. That just can't be true. Maybe you met the wrong people or something, but this is ludicrous."

"I wish, man, I fucking wish." Pau's eyes had no tears and no emotion. They never dilated or enlarged. She spoke without ever changing anything at all about her face or her tone. "Think of it like this. When the sun shines, is it always sunny outside? No, right? It could be sunny in the middle of a tsunami or earthquake or something. What's my point? People can smile to your face but have a heart of coal with the worst intentions. And I know those people."

"So, you shut down people without letting them in by making your life a mystery? A little mystery is good, but not *all* mystery. You're really dope and all, but don't do that."

"You know, I never heard someone actually call me dope."

Timidly I responded, "Well, I'm glad to be your first."

"I may not have my life together right now, but it doesn't change a thing. People are still people no matter what stage of life someone is in. Eventually I'll be at a place where I can provide for my brother and give him a solid place to grow up. It'll be cool, because we'll be in the West Coast with sun and no more snow or cold and surrounded by all the people who smile and not have any bad intentions."

After her impromptu speech on how people suck, Pau gave her final thought on the poem, "The Frost dude was wrong. He really should've said the road to your destination isn't always taken by everyone, but the destination should never waver. For me, somehow some way with my mom watching over me, I'll create a family that I never really had."

And just like that, the conversation ended, and the plane touched down at LAX. Before I knew it, we were walking down the aisle and in front of gate A16.

"Hopefully, we'll meet again, Foster."

"That's a bet, Paula Middle-name-I-forgot." She smiled. Earlier I told her my name was Foster Macario Lidden. The initials spelled FML. And with that, my brother and I went towards the baggage claim while she went the wrong way. I would've stopped her, but it seemed like a lot of work.

Our bags were delayed and eventually I ran into Pau one last time. We talked as time passed while we waited for our bags to show up on the conveyor belt. We exchanged Twitter handles,

Snapchat username, and Instagram accounts.

"So, am I ever going to see you again?"

"Maybe. Haven't decided yet. Depends on how I'm feeling."

After her final words, Pau grabbed her black military-like satchel and rose gold suitcase off the belt and walked out of the sliding doors and onto the street. She hailed a taxi instead of an Uber. Her rationale consisted something about how Uber just came out of nowhere and disrupted the whole industry and now people are losing their jobs.

What happens next and is it a happy ending?

The flight from Chicago to the city of angels was the only time I saw Pau. I traveled across the country to spend time with family. She traveled to escape and make a destination reality even if the road traveled was unconventional.

Later that night we snapped and said she found a place to crash for at least a week. She even told me the neighborhood she was living in. Like anybody who knows nothing about a city or a neighborhood, I did a little research to find out more. From my amazing detective skills, I surmised that the area wasn't the best. I knew rough areas and I knew she was smart enough to avoid trouble, but this time it was different. I was rooting for her, I really wanted her to make her dream a reality, so I was already on edge.

We snapped, sent funny things on Twitter, and liked each other's post on Instagram for the following days. But one day, she just went completely silent on everything. I didn't think too much about it. I figured it was her keeping her mysteriousness, and if she resurfaced, she would always know I'm there if she needed anything. Still, something didn't feel right. Regardless, I didn't let that affect my trip and my purpose for being there.

On my final day in Los Angeles, I decided to read one of the online papers. On this lucky day I just so happened to read an article that was about a fire in the neighborhood Pau was living. The article detailed that an apartment complex was burned in a devastating fire. I wouldn't have cared, but the article went on to detail a girl in her teens jumped out the window from the fourth floor to avoid the possibility of burning alive. I only kept reading because the girl was described as wearing a silver bracelet with flowers. The article never said the type of flowers, if there was a clasp, or if the girl was identified.

It was a somber day. I was leaving my family who I wouldn't

see for months, maybe even years, and I thought Pau was possibly dead. I couldn't do anything about either situation.

Then I thought about it. If Pau is half the person that I thought I met, then I have to believe that jumping from an apartment to avoid a fire wouldn't be the way she would go. Pau would look at the fire, laugh, then walk through it and out the building. That's who Pau is. It didn't hurt that I prayed for her every night in LA. I know one day she'll resurface, maybe a different name and different road but the same destination.

LESSON 2:

The best oxymoron: now and never

My mother always said black boys were like an endangered species.

I disagree. Endangered species are protected and cared for not hunted down and shot.

I think it's worth the chase

I've never truly thought about it
Hallmark doesn't really make movies for my type of people
So it's hard to fantasize about it
I figured it would happen
In a spontaneous way, like out of nowhere
and then boom
the birds would stop chirping
and the wind would stop blowing.
You know I wanted that fairy tale ending, where
the unlikely, unsung, unmotivated man becomes
a euphemism of the universe showing unrighteous and
unparalleled justice.
I wanted something tangible
but invisible,
Something not easily divisible
but dependable.
Not some somber scene where Shakespeare shakes off the dirt
and
takes a seat in the audience while he writes notes
and claims that this is a true tragedy, Hamlet can't compare.
Yet, seeking rational approval is truly abysmal
since we all experience it differently.
I don't like alluding to it, to this,
it's a scary thought filled with fright and frostbite
Some people like me would hope that this event of euphoria
will bring them something everlasting
I don't know anymore,
I'm pretty sure that it was supposed to happen by now.
The question was: do you remember the first time your partner
said they love you?
Followed up by: do you remember what it felt like?
I stopped,
I was speechless
I'm tired of trying to deny it, no more lies,
Time to accept the truth.
Time to grow up and realize that maybe my fantasy isn't meant
for me.
I used to struggle with accepting it,

but now I've come to peace with it.
I know what I'm waiting for
One day that fantasy, the one I imagined, might come true
I don't know a lot of things, except that I won't settle for
anything less than that feeling
And that feeling is what I'm chasing,
I'll find it, possibly, maybe, hopefully,
breaking a wishbone in half with the girl that I love
but she will never see me as anything other than her best
shoulder to cry
and rely upon and making a wish under a shooting star with a
starry night sitting
underneath *the* residential giant tree,
I don't know but I'll feel the rush
Not of adrenaline
but the rush of the toxins that I was taught as a boy to repress.
I don't want to waste my breath anymore
I want to feel it.

Modern English sonnet (music)

Whether black or poor
Filthy rich or simply American white
There is one thing that transcends race and it isn't lore
But tell someone that this is music and their passions ignite

The passions they didn't know they possessed start to burn
You don't have to be from a small town to enjoy a song about
    community at a bar
You don't need to have heartbreak from your ex to see what was
learned
Music transcends, it bends, it doesn't spend an entire lifetime
    wishing and staring at the stars

Music has different genres and reaches to lifestyles and
emotions not skin tones nor ethnicities
It may be true that certain racial groups created a particular
    rhythm and sound
But it doesn't mean that sounds should become monopolies
With music you should be proud, no one should tell you can't
    listen to it because you're brown

Music tells someone else's story and at times can make one look
    inward for self-imploring
Question this: What if we all had the same life stories, wouldn't
that be boring?

If I was a rapper and wanted a banger

What to do if you wanted to make a song that was a banger.
Here are the steps you need to take.

1. Find a nice beat that is heavy on the treble. The beat is nice enough where people can hum it when they are walking somewhere with people.
2. Add a classical instrument to the beat; piano, cello, or flute are highly recommended.
3. Once the classical instrument is added, slowly add it and make it more prevalent. By the end of the song, the classical instrument adds flair and spunk.
4. Listen to the completed track and choose a spot in the beginning to start rapping
5. The first two lines being spit on the track have to rhyme; if you don't rhyme then you have to have clever wordplay with someone/something relatively know or a nice hook
    1. For example: She's no Robin but she's thick
    2. Or: I thought I won, but I lost and what I lost I never had
6. Now that you have the listener with wordplay, rhyming, or catchy hook/chorus combo you can continue with the piece.

(Steps 7–9 are in no particular order)

7. To create a banger, you have to reference a sport and an icon within that sport and then hit the listener with a bar that rhymes.
    a. For example: I'm a shooter so they call me Kerr, I pull up with all this ice and your girl is like brr
8. This step is hard to follow because there are so many outcomes. You must make a comparison between yourself and either an animal or a very luxurious vehicle.
    a. For example: she thinks that I'm a cheater but I'm as fast as a cheetah so when she leave, I don't even need her, Ima go fuck this other hoe when I'm smoking this reefer.

b. Or: she asks me why I finish fast. I told her I do like my cars do. This Maserati go vroom vroom, so she only gets the boom boom

9. A banger needs to also be dedicated to a family or people who got you to where you are now or want to be in the future. The bars used in this step are heavy on phonetics and it does not matter if they rhyme. Your best bet is to aim for clever wordplay.

   a. For example: my daddy in heaven so you know I'm getting the blessings from above. Above so I stay away from the beef I float like a dove.

10. After steps 7–9 are completed and added to your rap, you now have creative flow to write anything you want. This is the part of the rap where rhyming and puns are used to fill time.

11. Once you finish writing the time fillers. Go through your track and drop the entire beat and have a five-second clip, then the beat resumes with any bar from 7–9.

12. Whatever bars you didn't use in step 11, throw them in the song somewhere. Where the bars are located varies by each artist and purpose of the song.

13. Listen to the whole song. It sounds good, but it's not a banger yet. Go back and add the ad libs. The ad libs are what the people remember the most. Add the ad libs at the end of a bar that you think is really fire.

14. Now you have completed the track, send it out into the world. Sit back and watch your fame and wealth stack.

If I wanted a banger: rap

(This is a collaboration piece where I am the featured artist. The
other artist is also an aspiring artist.)

Oh wait. . . stop and pause
never bleed like it's menopause
Respect my name but call me Dennis cuz on these tracks I am a
menace
Fuck that shit
Fuck yo shit
I hop from girl to girl so the crew call me Tigger
Fingers don't gotta be up on that trigger
Cuz I roll with the killers who only get high off them meds
And them boys you be seeing we putting them in beds (shh)
Sorry to say yo ass in a box
You getting clapped, go down to the docks
They see the target and when they shoot they be calling Kobe
Cuz that boy's gonna be without a head from a body (ah)
The life I be living be having me sinning
Sad to say not all my boys see me winning (aye)
So this is for the boys that be resting easy,
Justknowimagocuanddoitforyouandmakesureyomamammagot
    protectionfortwo (oomm)
With us that chopper go ratatata and we blast off the ops
That boy over there, yes he is a cop (um)
so we see him and all you be hearing is pops (rah)
My life go crazy, my life go stupid, my life be fast I went from
    eleven to thirteen
cuz fucktwelve fucktwelve fucktwelve
If them boys wanna snitch we send a switch and put the body
    on shelves (ah)
I'm getting this money it ain't in no hurry
I'm paying them bills so mama don't worry
I don't play football but trust me I ain't fumblin', I'm rumblin',
I'm tumblin'
My bag secured, the guns are loaded, this bitch is bad, money
keep running in
I'm running to the top and ain't goin' stop
If you run up on me, you gonna get shot in the top (rah aw)

Aye. . . aye. . . aye. . .
(Chorus of *aye* repeating in the background as I talk over the
track)
Young K in this bihh with (other artist)
You know I had to do it to them
This for the whole crew
(producer name) we killed that shit mane'
We gonna prosper, we gonna succeed, can't go back
Psalm 46, I know God looking over us and watching
(ugly chuckle in the background)
And we out

Questions

Let me ask a question:
If a tree falls in the forest, does it make a sound if no one hears
    it?

So, if a boy falls on the street and no one sees him cry, is his
    masculinity still intact?
If the boy goes home and tells his father,
does that count as being strong for admitting his failure?
Or is it seen as weakness for succumbing to temptations and
    telling someone he failed?
Is his acceptance of failure something that the father should
    accept for raising a son?
Will the father take the time to tell the boy that weakness
    should be killed,
    like emotions?
Is the boy allowed to make his own version of masculinity?
Is he even allowed to question what masculinity is and the
    purpose it serves?
Years passed from the day where the boy fell;
is he allowed to wonder if things would be different if
he wasn't taught to believe that being a man means being
rough, being tough, or trusting your own abilities?
His body is now scraped with scars some more recently than
    others,
but does he tell people that he is in pain?
Physical and mental?
What would his father think?
Would his father approve?
Does his mother know that her son is being raised
to be mindless and without emotions?
Does his mother know that all the time she spent
driving the boy to games that he would rather be sleeping?
Do the parents know that the boy struggles with only liking and
    focusing on girls?
At a party late at night when the boy's friends say it's too early
    to leave but
the boy's body says he's already late, who will the boy listen to?
At a party late at night when the boy is past his limits

will he tell someone that he might not be able to drive home?
When the event is shut down and the boy decides to finally go
    home,
will he ask someone to take him instead?
On the way back home
why does the boy decide to switch lanes so that he can see the
    bright lights?
What time were the boy's parents notified they should meet
    their son in the hospital?
The parents arrive and see the boy in pain with a faint
    heartbeat and only ask;
        will he live?
The sheriff walks into the waiting room and asks
which parent is responsible for letting the boy be irresponsible?
The doctor walks into the waiting room,
he asks if both parents want to hear the bad news in private?

A letter

Why don't the people we trust tell us children stories only exist
    for children?
What I'm about to say may sound ridiculous but to me it's the
    opposite of ludicrous
I was raised on fairy tales and godparents, children shows,
    Disney princesses and all
I was raised on a magic school bus where Tigger and Pooh Bear
    were the best of buds
where adventure in the backyards with friends would end with
    a snack,
where girls were more than just power and puff,
and wonderful wonder pets could help a baby animal and save
    the day,
on this magic school bus, a bald baby boy told me not be afraid,
where an anthropogenic anteater taught me how to deal with
    Benny the Bully.
I wish somewhere on this magic school bus Mrs. Frizzle
    would've told me
that these shows have an age limit and are only meant to be
    figurative not literal
Yet I wished and I wished with all my heart to fly in a land of
    dragons far apart
I wish I didn't have to find out this way
That fairy tales and children shows ruined my life at a very
    young age
So let me carefully craft a letter of recognition on behalf of my
    contempt
I was allowed to believe in faith trust and pixie dust and that
    was a complete bust
So now I look back and ask whether I loved these shows or was
    it just lust
So Dear *Backyardigans, Teletubbies, Sesame Street*
Why did you give me high hopes?
I said I want to make the world a better place not end up
    looking for some saving grace
But amazing grace how sweetly I found that lies are told to the
    young children's brain
Hate to say but these lies may cause some pain

And the pain brings up childhood memories that come every
     second like non-stop delivery
If a tree falls in a forest can you hear that?
If you can hear that then let me again wonderrrr,
how much wood could a woodchuck chuck if a woodchuck
     could chuck wood?
I'm now realizing that woodchucks don't chuck wood
and peter probably never picked a peck of peppers
and sadly sally never sold seashells by the seashore and
You know New York, you need New York, you know you need
     unique New York,
but New York's not unique, it's not Chicago, it's not nothing
never minding that notion although it bodies an ocean
but it's not home, correction it's not
HOMES (Huron, Ontario, Michigan, Eerie, Superior)
So here I am telling you that I trusted tongue twisters,
fairy tales and children shows
But the jokes on me
cuz I never thought this is how life would or could be
To all the shows I mistrusted,
did you know that the world isn't what you taught me to
     believe?
So please, for future fans out there,
Tell them the truth
so there's not another me out there with the youth
I'm not really big so tell the next Sean out there that one man
     can change the world,
but he's gonna need some help
Tell them something,
Something that is a reality and not a farfetched fantasy.
Sincerely,
Your biggest fan

The Gap

What's so special about a scene where a best friend casually picks up another best friend from the airport? It's been nearly twenty-five months since Danny and Ira last saw each other. Danny was on a service trip in Fiji as a volunteer for an organization that offers medical support to areas ravished by natural disaster and neglected by an incompetent government.

The trip was dangerous, but even worse was the wi-fi connection. The two could barely string enough data to text weekly let alone FaceTime once every blue moon. What's more fascinating was that Ira was picking Danny up from the airport. Since adolescence Ira hated the idea of driving a car and never cared if she went places. She also vehemently opposed the idea of driving since in some countries women are not allowed the right to even obtain a license. The car being used to pick up Danny did not belong to one of Ira's friends, it was her car that she purchased with her hard-earned money as a barista.

Being a barista was hard work and being a former cheerleader whose income was as a barista was tougher than that. Ira didn't worry about the fact that working at a hipster café she was paid below minimum wage. Why wasn't she worried? Tips covered the difference and often involved her being paid close to $22 per hour. Since she was naturally every man's preferred height (5'3" and a quarter to 5'5" and a half—Ira was 5'4") and had freckles that showed when someone complimented her, then why would she worry about a thing?

One night when she was closing the restaurant, a man ran across traffic just to say she was beautiful and that her curly hair was astonishing. He eventually asked for her number, but she kindly said no. This guy wasn't better or worse than anyone else who had hit on her; Ira was just waiting for and waiting on someone else. She had already found her Mr. Right and was positive that these men were all different versions of Mr. Wrong.

Twenty-five months ago

So why would a former cheerleader with a major in gender studies and political science turn down the opportunity for love? Well, the answer to that isn't as simple as one would hope. This is the cliché/corny/predictable part of the story: Danny becomes hugely important after he leaves. What did one male best friend say to his female counterpart? If someone guessed that he told her he loved her and had since the moment they met on that track after high school, that person would be right. If the person also assumed that the two were sitting down and having a conversation at Panera; if that person assumed any of that, that person would be right.

Danny proclaimed that he wished every guy Ira were with was him instead. He admitted that whenever she sent risqué photos to him first for approval, it made things even worse. The worst thing was that he told her that this service trip wasn't also going to be doing good for other people only. It was going to be a chance to let his feelings for her die and allow himself to move on.

If Ira had to describe her best friend, she would say that Danny was an above average looking guy with no noticeable features besides having two different colored eyes. Ira would then say that Danny may not possess a sharp chin or luscious lips, but his witty charm and unique personality made up for everything he didn't have.

When Danny told Ira all of this, Ira didn't know what to do. Some could say that this was an information overload, the author of the story would say that this is called extravagant detail.

Ira never realized why Danny never committed to relationships until now. Danny had flings here and there and sometimes a relationship blossomed for maybe two months, but nothing more than that. Unlike Ira, who would talk to different guys at the same time to find out which boy to make hers. Those relationships often lasted six months or more and involved the words of "I love you" very early on. Danny was still her best friend, but she always figured he just cared about her like a sister. No one can blame Ira for confusing affection as protection, especially if the protection was coming from Danny.

Danny lost his only sibling, a sister, in a plane crash coming home from a summer service trip to Africa. Danny's sister volunteered through the local Catholic parish, the same one Danny attended monthly since his confirmation and communion. After the funeral for Danny's sister, Danny vowed to never do a volunteer service trip through the Catholic Church. And Ira agreed that she would never allow Danny to do the same thing (part of her Protestant side came out every time religion and Catholicism was mentioned), because the thought of losing Danny forever was worse than Hell.

Putting all this together in the short time, Ira had to gather her thoughts and choose what to say next, but she couldn't utter a single sound. She was speechless and at a loss for words. Like a boulder smashing an infant, this hurt Ira. Danny's sudden announcement hurt a lot more than she realized. She tried to protest Danny's leaving but no sounds came out of her mouth when she opened her thin lips. Ira only thought about the vow she made and how she wouldn't be able to keep it.

When Danny mentioned that he was going on the service trip it took Ira aback. Then when he mentioned that he was leaving that day for two years, it made her eyes swell with water, the same hazel eyes which the puppy dog eyes are named after. To make matters worse, Danny said that his Uber was waiting outside to take him to the airport and was already running late.

Ira was given so much information to take on with no time to process any of it. At the end of Danny's speech talking about what he was going to do and his feelings for Ira, he did what all people who live in movies do. He looked into Ira's hazel eyes, grabbed her hands, and told her that these will be the toughest two years of his life and something along the lines that he will never love someone the same as he loved her. And with that, Danny let go of Ira's hands and walked briskly to the Uber. On the way to the Uber, Ira noticed that Danny failed to comb his black hair; evidence of the way he was pushing his hair aside from his multicolored eyes.

Ira was left alone at a table for two at Panera. When the Uber pulled off, tears streamed down her face and they didn't stop. No one noticed the tears stream down her adorable roundish face because she was seated in the corner. Whenever Ira and Danny went anywhere, Danny always found a spot in the corner. Danny's views on why he chose to sit in the corners always fascinated her. Matter of fact, most of Danny's views fascinated Ira; Ira found Danny's views on life refreshing. It was one of the many things she loved about Danny.

Besides feeling every emotion, Ira felt the emotion of rage the most. Who texts someone to meet them at *their* Panera and then drop bombshell after bombshell and then leave? Not just leave and see you later, but leave the region, country, and hemisphere for two years. What pissed her off even more was that Danny saw her order food and then proceeded to order food as well. Who does that?

As her rage for Danny bubbled, one of the servers at Panera delivered two plates of food to a table that only had one person. Ira hid her face in one of the many hoodies she had stolen from Danny so that the server would not see her crying.

Ira's order was a simple panini with chips. Danny's order was barbeque mac and cheese with a side order of a Caesar salad. Ira thought of how much of an asshole Danny was. Danny knew Ira hated the idea of boxing food up and wasting the immediate flavor that the chefs so beautifully created. So, she began to ponder as to why Danny would order her favorite food from Panera then leave.

Ira stared at the two plates and began grinning. She realized that Danny planned all of this. That asshole, she thought. That sly dog, she thought. He was forcing her to enjoy food even after a traumatic event. In a mini flashback montage, Ira remembered all the times that Danny forced her to eat: after rough breakups, failed midterms, and being denied from any job that required someone of her major. She remembered the laughs they had where Danny would say, "Life comes and goes, but food will never betray you" (a lie because the two had received food poisoning from some Thai place uptown once). Danny's thoughtfulness and attention to detail was one of the reasons she loved him.

Ira never thought she could imagine the day that a novel written by Nicholas Sparks would come true and she would be the main character. But there she was, crying alone in Panera with a panini in one hand and a mouth full of barbecue mac and cheese. As she stuffed her face with food, she took her headphones out of her bag and put them in her ears. She didn't think and hit the first song on a random playlist on her music app.

For a brief time, the tears appeared to be coming to an end. Moments later it appeared that the tears decided to not stop and continued. The difference was in the tears. The music being played was a song Danny showed her and was now one of Ira's favorite songs. Tears now fell down her face and onto her phone where she tried to wipe them off the touchscreen with her stolen hoodie. The playlist was custom curated by Danny for Ira and was updated weekly. If Ira needed a boost, or needed time to relax, or for any occasion she would listen to this playlist. The tears weren't of pain, sadness, and anger; the tears were caused by happiness and joy. Danny knew her so well and was always there for her. Another note to his amazing attention to detail and how connected the two really were.

Ira knew when Danny let go of her hand that she would miss the ever-loving piss out of that scrawny boy. She would miss his goofy smile, his nonchalant hair flips, his nervous tick where his face twitches uncontrollably on the right, his values and beliefs, his need to make sure she was safe and okay, his uncontrollable way to make a situation better, and his energy. Of course, she thought she would miss her best friend and that is completely justified. What she didn't realize was that she described the feelings and actions of someone who was in love, or in this case, denial.

She loved Danny like all best friends do, but now she understood that she did in fact love Danny the same way he loved her (not the same, because love is different even when people are in love). She loved Danny not like her best friend, but like he was her Mr. Right. It only took this slender lanky excuse of a boy professing his love for Ira and then telling her he was leaving for two years to allow Ira to figure it out.

Present day

So why was today so special? It's just one best friend—the same best friend who loves the other best friend but never told that person how she feels—picking up the other best friend, who professed his love for her and then left for twenty-five months.

Ira had been looking for a spot to park the car in the arrival lane of the airport for the past ten minutes. Checking both shoulders, she snuck her car into a tight spot while a bigger Toyota decided to blast their horn at her for cutting them off. Ira shifted the car in park and breathed a deep sigh of relief.

Today was the day she was going to see Danny again. Today was also the same day she decided to tell him how she felt. All the times they texted or FaceTimed and she still didn't think to tell him. She told herself that if she was going to do this, she was going to wait until they were face to face.

Just like Danny, she had a plan. She would pick him up and take him to Panera. *Their* Panera. Along the way she would ask him what he's been doing and hear his stories from Fiji. At *their* Panera, she would order a panini and hold his hand then tell him she loved him and then leave. She wanted to leave for five minutes for the drama. She had it all planned out. She knew it was going to work out. She just knew it.

Five minutes after securing the parking spot, Ira received a text from Danny. Ira's heart jumped out of her chest as she assumed the text was saying his plane landed. The last time the two spoke it was about a week ago and Ira was positive nothing had changed. They would have talked more, but Danny warned her that his service wasn't going to be reliable. Ira's phone vibrated again. Like any sane person, Ira lowered the volume of the aux in the car so that she could better read the texts from Danny. Her phone buzzed again. Ira finally pulled the phone from its location in one of the cupholders. Ira turned the phone over and read the texts.

Picture

"Sorry to drop this on you last minute, service is really shitty. That's Joanne. We met on the trip and we really connected. I know I've never mentioned her to you, but yeah that's her. She's beautiful and amazing and just wowww. While helping others we found a connection and fell in love. I love her. I wish you could see how her smile lights up the room or how her love for life makes me wanna change the world. Long story short, she was going to come back with me and we were gonna live together but things took a turn. Joanne is pregnant. The kid's mine, don't worry.

"We found out two days ago. We're scared, but super pumped. Like pumped enough to touch the stars. We were just joking with the idea of raising a kid in a different country and on the values of helping those in need. The jokes became serious and we decided to work full time for the organization. They agreed to put us on the permanent payroll and we're going to set up at the HQ for a while to start our family.

"When the kid is maybe two or three, we'll move again, but until then I'm still going to be on this side of the world. There's no easy way to say this, but we'll FaceTime soon, and I'll show you Joanne so you can meet. The crazy thing is that I've only known her for less than two months, but I know my life isn't complete unless she's there.

"She reminds me a lot of you."

Ira never got to tell Danny how she felt. She never got to hold his hand. On this day she wouldn't even see Danny's smile or Danny's hair. She couldn't tell Danny.

LESSON 3:

Acceptance is harder than rejection

People complain that they're not perfect.

But no one asked them to be.

What everyone thinks when they fall in love

High and lows
Eyes and nose
Xs and Os
What are those?
Ups and downs
Mostly upside-down frowns
Twisted from ear to ear
Intentions are clear
What are those?
I feel that I froze
Oh look: more Xs and Os
I wonder if my passion shows
Is this a dream, I think or as I doze
What are those?
I feel lighter
My smile becomes brighter
And the situation is dire
What is this?
I'm not good at talking
Barely okay at walking
Adequate at balking
What is this?
Being around you I think I'm always skipping beats
Happy feet,
got me feeling that I need to take a seat
If this is a game, then you are a cheat
What is this?
Actions speak louder than words
I remember because I want you to know you're being heard
Next to you I can't see straight, everything is kinda blurred
Honestly, that's quite preferred
because I don't want anything to be unheard
What is this?
Jordin Sparks asked how is she supposed to breathe with no air
But if you asked me,
I'd always carry an oxygen tank in my bag with room to spare
So, I could also hold your heart
Have you ever seen something and marveled?

If you haven't,
you should just look at the reflection in the mirror that stares
back

Modern limericks (social media)

They say it didn't happen unless you post it
And everyone wishes it's not true, but it's not bullshit.
This modern generation is something unique.
Instead of helping, they complain that everything reeks.
This modern generation is the definition of quit.

————

To my surprise a friend once tried to do something nice,
They said take a video and if this blows up, they'll be living in
    paradise.
What is genuinely wrong with people?
If this is how people think, then I need to jump from a steeple.
Some people eat full five-star meals then do something nice
    and post a pic handing out rice.

————

People who are only nice online really grind my gears,
They deserve to go back to the age where it was just bows and
    spears.
If you're going to post something good just because you want to
    get notifications,
I hope you only gain notoriety from everyone in your nation.
To the people who do good deeds because they're good
    humans, cheers.

Modern relationships

I write a letter to lively lovers that listen
Not essays for educators to endlessly evaluate
I speak soft soliloquies of
silk syllables
so stay,
stop,
share the snore filled slumbers by my side
But it all begins with a
wishbone whisper wishful
wonder thinking
tantalizing
turn twisting text
"yo. . . you up"

Mene mene tekel upharsin

To stare at the moons and stars and know that heaven is this
    close
To smell the salt water in the air and know that escape to the
    sea are steps away
To listen to nature speak while standing in the quiet valley
To feel the jagged edges of the rocks and know the weather
    wanted wonders
To savor the aroma of life and know that life is not as sweet

Mene mene tekel upharsin, to be judged and found wanting
Wanting as in not worthy and wanting as in not good enough
If God exists,
hands of fate, or whatever controls the universe
something has to know that humans will never be worthy
There are sins we can't erase that are passed down from
    generations like diseases
If we were to compare ourselves to people of the Bible,
God's "chosen people"
we will never be "worthy,"
we'll only be considered unholy no need for phony stories or
    matrimony

Cain showed us that anyone is able to do anything
Towers have been made that rival Babel,
If there's a modern-day Jacob who has thirteen sons,
understand none of them will buy him a multicolored shirt
and deliver a savior born from a manger and
promise to deliver salvation to strangers
The way people fall down they might as well be Jericho
hoping that trumpets don't sound when they already gave up
    days ago
Everyone's concerned with going to a land of milk and honey,
when people ought to wonder if they can afford the trip there
    which is funny
Goliath is now a figure of speech when something seems
    impossible to beat
so we're all different forms of David when the challenge
    becomes a feat

Solomon split a baby in half so no one could be happy
but when we disagree with a statement, we're considered
        unworthy by our peers
and that is the true definition of living in fear

As a collective,
we will never be the people from the sacred texts;
And to compare us to the people who only worried about their
        beliefs
and not finding social relief
that would make us unworthy of the world which we inherited
        like estates
If the world had Israelites posting selfies while wandering the
        deserts
or photos of Ezekiel ascending to heaven,
the world would be different

mene mene tekel upharsin,
we have been judged and been found wanting
We may never be worthy for what we have,
but that doesn't mean we have to stop,
Stop being who we are or
stop trying for what we want

I know, verbatim

"You think I run in a gang," said the white boy who believes that he should be able to say the *nigger* word because he hangs around colored people and went through 'a lot' as a child. "If you have something to say, say it to my face. Keep my name out of your mouth. None of this sneaking shit unless we have a problem," he says through text.

I don't know if I was supposed to be scared or threatened, but I began to laugh. Does he know? I don't think he knows. I'll explain. There is one person in this entire universe that scares me and makes me rethink my actions and words. For my sake, I don't see them often enough to understand what fear is.

I gathered my thoughts and responded, "Don't you know where I live? Right? You've been here multiple times. I am so sorry that you became involved in a conversation and situation where you need not to be involved in. We're not on personal terms but I feel that I should tell you something, three things to be specific. First, where I come from, I was taught how to properly operate a gun: from proper stance to lining up the target. Second, I don't know karate, Thai, wrestling, or anything else. I was taught simply how to whoop a nigga's ass and make sure they can't walk when they're down. The term 'on their necks' was one of the first phrases I said. Lastly, if I want to talk shit, I will go up to you and say it to your face. It won't matter if your mother, friends, girlfriend, pastor, or Jesus himself was next to you. I will say it and not back down one iota or lose our intense staring contest until you realize that I'm bat shit crazy and will shove my shoe down your throat."

I placed the phone down and then immediately picked it up again. I saw that he was trying to respond, evidence of the three dots (thank you, Apple). I decided to respond before he could, "I will put whatever I want in my mouth, and that includes your name. I mean I don't blame your mom when she was constantly screaming my name."

I think he was mad. He responded. He responded again. And again. I didn't open them. There was no need, I made my point very clear and if he had a problem he knew where to find me.

"I hate when you describe people like that. Especially him, he's one of my good friends," she says. She is referencing the fact that instead of calling the raging and fake tough white boy by his name, I call him for what he is: a white boy who believes that he gives off the vibe of someone that people shouldn't fuck with.

I told her that it doesn't matter who you are and how much respect I have or don't have for you. I just preferred to call people by what and how I saw them. The way I addressed people is not derogatory or demeaning. It should never be considered a problem. I told this to the brown skinned girl who believes that a problem is not her fault if she caused it.

She believed that life was testing her by making her have feelings for someone like me. She told me Superman found a way to overcome kryptonite and so will she. I told her that there is more than one type of kryptonite. Her face disapproved of my comment.

She was sitting on my broken futon with her legs crossed and wore a winter coat that was zipped to the top although it was 73 degrees in my room. I don't like the cold and prefer the artificial warmth created by heating units.

Why was she here? The girl didn't live in this area of the city. I knew the reason why she was fully covered: she wanted to leave my apartment. Leave my room that belonged to my friend's apartment, and not be tempted to stay the night, like every other time we have conversations that go through midnight.

"Well that's really tough," I said with a grin. "This is my life and that's why I am allowed to say whatever I want." Her brow inched higher and her body made the notion that she was leaving, she stayed. This girl was not good at hiding her emotions from her face and showing the world. She was annoyed.

"This is why it didn't work out the first time and here I am again trying to just understand who you are so I can get closure," said the girl with large and supple brown eyes. Her eyes are hazel, but they have a tint of green somehow. She mentioned this fact about her a lot. You can tell that that's the feature she prefers to receive compliments on. I told her she looks cute when she's mad and that the way her lips curl up at the end remind me of nights when she cried from laughter. She was not pleased by my comment, evidence of her lips which curled even more than before.

"He doesn't have to go through what I went through or will ever have to. He's handsome and can get away with it," I told her. She saw me chuckle. She tensed up and was noticeably upset.

She knew that's never the point I would try to make. I regained my sense of self. "This is the hill you want to die on."

"No, I won't die. Just you always misjudge people and mislabel them. You place them into boxes, and you won't let them escape. You put me in the box of a damsel in distress. And as much as you want to, I didn't ask to be saved. I'm not even cute," said the girl with brown and a tint of green eyes. She was fidgeting on the futon; anyone could be able to tell she was becoming flustered.

"You don't need saving, of course not," I said sarcastically. I raised my voice slightly and shifted tones, "Let's not forget where you were a week ago. The people you surround yourself with don't care that you almost threw away your life. So, you may not need saving, but you need something, and hate to break it to you. But that's probably me." It's hard to be serious or taken seriously when sitting in a comfortable saucer chair, but I tried my best.

"And did you notice that I didn't call you, I called other people. I called him. There's a reason. They won't tell me about their disappointment in me, or how mad they are at me or anything you like to do. They would notice that I'm down and won't make me go lower." Her forehead got smaller as her straight brunette hair with blonde highlights fell into her line of vision. She grabbed the bag of cookies next to her and stuffed her cheeks full. There was a knock on my door as the conversation intensified.

I looked at her and snatched a cookie from her dinky hands and told her, "I don't give a damn if you're in Heaven. I will tell you how it is. I don't coddle, I cuddle and right now I'm doing neither." My face turned from serious to playful as hers shrank. The emotions of rage with disappointment were shown on her face. "You understand you're a shit stirrer. There was no problem. You caused the problem; you may be the problem. But you can't accept that and will toss blame to the nearest target. And guess who it is, me."

Her eyes swelled with water; she was still fidgeting on the futon. The cookies were placed to the side as she moved on to fruit snacks, "What do you want from me? Yeah, I know I messed up. I know. I know. You're not the only one who tells me this."

I was at a loss for words. If anyone knew me, then they knew I eventually said something, "What don't you want me to say? You are the first one to get blamed, and that's exactly what happened. This isn't funny. Your situation isn't funny, but it will be if you don't learn anything from it. If you don't try to grow from a moment that you are lucky to escape with your life, then I guess I am wasting my breath."

"I guess you are."

"I guess I am." I moved from the saucer chair and lay down on my shag carpet inspired by the '70s.

"I'll be fine," she promised. While fixing her hair, she began to tell a story about how she cut her hair for some asshole she was with. Her hair was now in somewhat of a bun. Whenever the two slept together, her hair was always in his face. She cared deeply about the boy and cut the hair so that he could have a better night's sleep. She says that he should've known she saw something in him and with him if she did that. She ended the story by saying she doesn't understand how the two are in a predicament.

That boy was me. The story was shocking. I didn't know she did that for my benefit. I wish she would have told me. I don't know if it would've changed the outcome for anything, but it would have been nice to know.

To play devil's advocate, she could just stop sleeping in my bed and avoid cutting her hair. In my defense, I was only kidding when I said the hair was bothersome. Regardless, it was a nice gesture and one that went unappreciated.

Her story came to an end and silence descended on the two of us. The silence remained as the casual sound of my roommates' talking became prominent. To end the silence the same handsome white boy texted me.

I turned the phone and showed the brown skinned girl with fruit snacks in her mouth. I got up from the shag carpet and came to a crisscross applesauce position to speak. "Look who it is. It's your bestie, a want to act tough white guy who probably believes he's a black man trapped in a white body."

"There we go again, why do you do that?"

"Do what, that? You're offended that I call it as I see it."

"But why? Why do you do that? Why do you have to demean someone else to make yourself feel okay?"

"Isn't that the reason why you're here again? To understand why I do the things I do."

"Yes, but you don't have to be a dick about it."

"I'm not a dick, I'm just unapologetically me. And I don't need to demean anyone to feel okay. I'm fine with who I am."

"Sure you are," the brown skin girl with the impressive eyes said. Her eyes were still filled with water, "but that's not the point. You can be yourself, but you can't act like everyone doesn't go through a struggle. And my friend, 'that white boy,' as you say, also went through a struggle. Maybe you should ask."

A look of confusion displayed on my face. "Ask, for what reason? I don't care enough. I met him because of you. Hung out with him because of you. And we're done, so I don't care enough about him. Like I said, I call it as I see it."

"Enlighten me. What do you see?"

"I see that as cool as you make him out to be, he will never, and I mean never have to go through what we have to go through. Especially me." I noticed that my tone was beginning to become shaky.

"That doesn't mean he's less than."

"That's my point. Your friend will never be less than. He will never be seen as less than. Never have a thought telling him that his skin will be less than. He will never be overlooked for a job, even if he doesn't have the qualifications."

"See you're putting him down by saying he doesn't have to jump the same hoops you do just to get to the same place."

"I'm not saying anything that's not true. Just saying things he will never consider, and honestly I don't see the problem with calling life and people by how I see it."

"Everyone struggles. Everyone has a different struggle. And for you to say that your struggles are deeper and more meaningful than anyone else is wrong. The same dumbass boy told me that everyone has a different interpretation of what it means when they say, 'life sucks.'" The cute girl uncrossed her legs and continued to eat my snacks in between her sentences. Her eyes had absorbed more water.

My legs grew tired from sitting crisscross style, and I switched positions to lay back down on the carpet. There was another knock on the door. "So, it's wrong for me to say—"

"Bitch, I wasn't done," said the brown skin girl with straight hair whose childhood taught her that if you're loud enough then you are allowed to talk. Her legs were uncrossed, and her coat was halfway unzipped. "I get it. Everyone who knows you, gets it. You struggled in life. You struggled a lot just to get here and you don't ever want to go back. That's awesome and all this rah rah. But that doesn't mean you have to describe people by what you think they don't have to go through. A girl could be privileged, but still be physically and emotionally abused. But all you would see is that she's not colored and never had to worry about whether or not she ever lived in fear of the same things that scare you."

She paused. Tears streamed down.

"Struggling in life is different for everyone. It shouldn't define who you are and how you act towards other people. You claim you're smart, but if you can't value that part about people then maybe you should rethink that."

"I am what I came from. I am what made me. Out of the fire, that's me." My voice cracked and it was noticeable. It broke the tension in the room. It was noticeable enough for me to pause and wait until I was positive it wouldn't happen again. "How you respond in face of adversity defines who you are as a person when it really matters."

She paused again to finish the last cookie in the bag. Mouth half full of cookie, she said, "And I agree. But how are you being faced with adversity in a conversation with me? Do I remind you of what happened for you to get the scar on your elbow? Do I remind you of the nights where you were terrified of your own shadow? If it's yes, I'll leave and never bother you again, but I know it's not." She climbed down from the broken futon and laid down on the floor to a place on the carpet where the tips of our hair touched. My unpicked nappy afro hair with a clean taper touched the tip of her tightly wound bun.

Tears still streamed down. "I'm not saying that it's not good for you to have a chip on your shoulder. But when that chip is all people see you as, then it's not a chip anymore. It's who you are. I know we're not on great terms and you're always talking to me like this, but you're more than a chip on your shoulder. Whether people had reliable meals or not. Whether you lived out of a car or in a condo, the struggle and life is different for everyone. There are a select few that may never experience any of that, and that isn't even dependent on social class. Your experiences and background shouldn't be the only things that make you stand out. No matter what or who you are, we all have different lives. From the places we called home and the memories in our heart, what defines us as a human being is how you carry on your life and what you gained from those moments."

In the middle of all this thought and reflection, a hand reached over and gently wiped the tears off my eyes. I was in no shape to talk.

"So please can you stop demeaning people because they

weren't raised the same and didn't have to go through what you did. It hurts me and in the long run I know it will hurt you." Seconds passed, she finished her monologue with a question, "So what are you thinking?"

I wasn't thinking. I had no idea I would end up in this state. I said what I knew, "I know that I don't know."

Her winter coat was left on the broken futon. Her sweater she wore underneath her coat was thrown on top of her coat. She switched positions. At this moment we lay next to each other. With one hand she wiped the rest of my tears from my red eyes, which were the same as hers. The other hand grabbed mine and gently rustled her fingers against my rough knuckles. We stayed in this position for several minutes. The minutes seemed like an eternity. We were surrounded in silence and lights from the mini star constellation projection lamp on the floor.

The automatic timer went off for the constellation projection lamp, and all that remained was darkness. She moved closer and I stared into her weirdly colored eyes. I was finally able to muster words, "Are you high?"

She crooked her face and responded, "As a kite descending from the stars. You're out of cookies by the way." And with that she tapped my forehead and quickly arose from her position. She then proceeded to get onto my bed. Her socks and pants were tossed into my overflowing dirty clothes hampers. She snuggled into her usual position and already possessed the majority of the covers. Only thing missing from this night; me next to her.

Time passed and I remained motionless. I laid there on the floor as the tears left white salt streaked residue stains on me as their parting gift. The bed shifted and a voice under the covers was heard, "So are you sleeping on the floor tonight?"

When the question finished, I stood up. I made the short trip to the broken futon and laid down. I used her sweater and winter coat as my cover. I didn't respond to her question. It appeared that the futon and I shared similar features: both of us weren't good at comforting people, and we were both very broken.

LESSON 4:

Life is what you make it

Black ice is scary, but black skin has been and always will be beautiful.
I just hope that one day we can understand the difference and draw a line instead of letting the two ideas overlap.

Is this love

I never liked the analogy or euphemism
it's like rubbing salt into a wound
An open yet healing wound
It's more like chainsaw in the neck on full throttle
No, no, no, never mind, that's not right
It's inaccurate and anaphora analyzing alternative aliases
It's more like boulder falling on head
falling into ocean
into a dormant undersea volcano that just happens to erupt
Spitting rocks and my skin into the sky
as the sun absorbs my soul and shoots me out to the stars
where they mimic monolithic black holes
that break my body and bones into cosmic dust
denoting my DNA as abysmal and deeply dismal
I think I'm grasping at straws,
but I think that's what it's supposed to feel like when I know I'm
    in love
I'm just asking for a friend.
So, tell me, am I wrong?
. . .
Yes, they say
. . .
They say love is indescribable
. . .
They say being in love is unimaginable
. . .
I shrug.
In that case,
I never knew love and love will never know me.

Modern day villanelle (tap water)

Don't play dead and try to make your eyes more round
You know exactly what you said
But remember I never asked you to judge my background

You screamed when you saw me drink water that came from
the ground
The scream happened and then your eyes were full of dread
Yet you give me eyes that compare me to a lost and found

In your presence I made you smile, but in reality I guess I was
crowned a clown
I forgot you judged me when I said I was hungry and still went
to bed
But remember I never asked you to judge my background

The sympathy shown is only there because you realize you had
everything abound
I wish you were there when I couldn't afford Whole Foods
bread
Yet you give me eyes that compare me to a lost and found

I lived in a house while you stayed on a compound
I drink tap water and unlike you for fun I have never gone to
the mountains and sled
But remember I never asked you to judge my background

I can't apologize for not growing up with Fiji and Voss water
around
Or that locking doors came with the threat of them being
knocked down
Yet you give me eyes that compare me to a lost and found
But remember I never asked you to judge my background

My speech

A vote for Nichols
is a vote with change
Free, free, free, and more free
MLK said let freedom reign
So to the ladies and women, free is what I say
All tampons and pads should be completely free
I don't know how this would happen, but it needs to be
But if pads and 'pons are free then people without perfect vision
should also have access
Free glasses, monocles, contacts, and everything
but not to excess
Cuz just like women didn't ask to bleed and proceed
Other people didn't ask to be blind, blurry, or blurred
Without modern technology people like me would be left for
dead
maybe I would've been misled looking for bread accidentally
end up without a head
Hear me out, my middle name is Muhammed, so that means I
stay dead
unlike Jesus so don't ask me to please us
as you wonder that I try to seize your attention like a
concerned Caesar's seizures when stabbed by backstabbing
Brutus
but hands up please Brutus don't shoot us,
Brutus has a gun and a badge,
modern technology.
What about people with crooked teeth?
People didn't ask for a smile to be considered imperfect
People didn't ask for themselves to be called snaggle tooth,
or snap, crackle, and crooked,
but if it's free would you keep it?
From the father down to the daughter
Understand Flint still doesn't have clean water
But this is considered modern technology.
Modern technology and free are two things that should exist
hand in hand.
If we can put a man on the moon,

then why is it when someone demands change it's considered
too soon?
If we have megacities being built on the backs and bones of the
poor,
why can't we have affordable health care when you walk out
your front door?
If we can condemn terrorism,
then why do we condone subtle racism?
If it's possible to make fancy gadgets and gizmos,
why do we still have people hanging from the willows?
With all this technology people still believe having a darker
melanin turns you to a felon
We don't need to hit a restart
or rip your heart a part to know that it should beat equality
This world rotates and revolves and
yet nobody can dissolve the issues in queue,
We're not playing pool but
we need to be hitting the ball on cue
Before we can give away free pads or tampons,
and glasses or contacts or even fixing some broken teeth
We need to focus all of our modern technology on having your
skin not being a predetermined sin,
A vote for Nichols,
is a vote for change

Mansplaining

Pretty hurts and ugly kills but but but but beauty is repetition
seen as a competition including malnutrition which deceives
    your intuition
and cost half as much as your college tuition.
Some girls primarily plaster themselves with forever pink
    products
proclaiming they want to be plastic.
It's not all that great to be plastic ask barbie how long it took to
    become "fantastic"
Pretty hurts and ugly kills because how often can one hear
    someone say roses are red
violets are blue and that's why my favorite flowers are your two
    lips
but people don't think they're beautiful.
Because if you didn't know, ugly kills, pretty is painful
and time flies like an arrow and fruit flies like a banana.
Don't forget ugly steals, ugly feels,
and most importantly ugly takes pills to drown out the
    depression
which is caused by the obsession to fix every slight imperfection
Claiming that the clock takes too long
but luckily ugly doesn't own a firearm
so go ahead and sound the fire alarm
because if ugly puts people on the edge of self-harm then we
    have an epidemic
a pandemic a global problem that started off systemic but is
    now academic.
We forgot along the way that everyone is priceless,
now we think everyone belongs on a prize list.
So many people think that appearances only matter to women,
but what about the boys who think they're too big or thin?
The idea is so wrong that girls should only want a man whose
    dick is long.
Men are just as self-conscious,
but they don't say anything because they've been taught to
    ignore their feelings
so expressing concerns isn't even in our subconscious.
But I guess my idea is wrong,

all men need to want a woman who will shake her ass in a
     thong.
Blame this on human nature, because we never humanely
     nurtured ourselves,
we try not to split our souls into fractions
before we actually try to put our words into actions.
If you wear make up because you like it,
go ahead, go right ahead, don't let anyone stop you,
BUT I'm just saying if you're using this to make yourself
     prettier because you don't believe you are,
I'm so sorry that we failed you,
you don't need to make up for what people don't see.
Because apparently it got lost in translation that people are
     more than a
one-dimensional physical plane perceived by the pupils.
Funny how this is what people do on the road to riches and
     diamond rings:
sell their souls for a quarter of their worth;
they also sold their bodies to the rich as a free deal
but understand the rich don't enjoy free meals.
Most girls would still rather appease others,
so they hide behind masks of mascara.
Most boys would rather shoot a gun or a ball
instead of coming to terms that inside they feel small.
In the end I'm just another guy mansplaining.
Just heed my warning that for some people, this is their only
     truth:
pretty hurts, ugly kills, and beauty is bold

Pau's story, drip

Water drips from a leaky faucet
The mother bangs on the faucet to make it stop
but sheer will won't make it stop
The mother convinces herself that someone else will fix it
The girl, her daughter, begins to laugh
She laughs like the toddler she is,
never knowing that a leaky faucet would cause hardship
The faucet leaks, water continues to drip
The mother brings home a man, he is a handyman who is tall
and big
Rough around the edges like a rock you skip on the lake
The man says he will fix the faucet soon,
The daughter can now do math,
she can tell that something doesn't add up
The faucet leaks, water continues to drip
The mother has a son,
the man is overjoyed with pride and excitement
He always tells his son he would have left if he was a she
The mother is tired from cooking, cleaning, and being a mother
The man does not care,
he is still jagged like a rock
The girl hopes to find an ally in the brother who cannot speak
yet
The faucet leaks, water continues to drip
The mother is covered in black and blue
The man apologizes every time and says it won't happen again,
he says every time
The boy can now walk,
he is being pushed down a path that looks similar to his father's
The girl cries because she can't do anything,
she yells at the one thing that's always broken
The faucet leaks, water continues to drip
The mother is gone,
The man seems to not care,
The boy is being forced to choose between sister and father
The girl has now replaced her mother,
her skin is tattered with blue and black bruises
The girl begs the boy to not listen to the man,

she says grow up and fix the faucet
The faucet leaks, water continues to drip
The man is not so big anymore as he is round
He yells when the food is cold,
The girl slams the door and
walks away underneath the starry night
The faucet stops leaking, it is only for a few seconds
The boy sees his father and wishes for his mother
The boy grabs the tools that have never been used
and looks at the faucet
The faucet leaks,
water continues to drip

A Conversation Online

Welcome to Chatpad
User87random has entered the chat
showm3boobi3s has entered the chat
yaboi_Alex01 has entered the chat
showm3boobi3s is face calling the chat
showm3boob3s has ended the face call
showm3boobi3s has left the chat

yaboi_Alex01: permission to rant?

User87random: I don't see why not. Permission granted.

yaboi_Alex01: sorry. it's just i usually use this app to rant about my day, my shitty day.

User87random: What made today so shitty?

yaboi_Alex01: what didn't it do?

User87random: Hate to break it to you, but I won't know it's shitty unless you tell me.

yaboi_Alex01: you're totally right. where to start?

yaboi_Alex01: i had this test and i totally bombed it. i wouldn't be upset with it, but i studied so long and so hard for it. Ha, that's what she said. i was trying to finish the test but the teacher said times up and came and collected the papers. i can't wait to get to college, no more timed tests i heard. im only a junior tho. so i still have a year left of this bullshit. so that was the second class of the day. then my gf dumped me through text. not even thr—

User87random: Are you serious? You just gave away your age. You don't know me. I could be a stalker. I could hack into the network and find your ip address. From your ip address, find your actual address. Then I would come to your house and kill you.

yaboi_Alex01: —ough text. she handed me a note in our fourth period science class. the note said check ur email. so i checked my email. she shared with me a google docs. the docs said she was breaking up with me so that she could start dating some guy from this other high school.

yaboi_Alex01: do u want to kill me, because i will give u my address no joke. after this day i dont want to be alive right now. so would u like to kill me?

User87random: No, I don't want to kill you. I was just reminding you of the rules that you violated. I think you might need help.

yaboi_Alex01: well that's disappointing to hear that u dont want to kill me. i dont need help, i dont need anything.

User87random: You still told me that you're 16. I could be a moderator and suspend your account you didn't think about that.

yaboi_Alex01: i know you're not a moderator unless u would've finished the conversation then suspended me, not threaten me. and i didn't tell u i was 16. i said i was a junior but u inferred that i was 16. i guess u could feel my gemini energy all the way from ur nice house. and if u realized that i was a gemini that means that my birthday is late spring early summer.

User87random: I'm not answering your question, and I never said I had a nice house.

yaboi_Alex01: point is, you're either living alone, a troll, a liar, or someone who is in their teens like me. those are most people i met on here.

User87random: Welp, I'm neither. I'm a student in my second year of grad school.

yaboi_Alex01: that's dope. im alex.

User87random: You're NOT SUPPOSED to give out personal information.

yaboi_Alex01: i mean it's literally in my username, lolol, if u want my social security or something, go ahead take it, i won't need it. it starts with a 9.

yaboi_Alex01: so can i finish my rant college boi?

User87random: Proceed.

yaboi_Alex01: so after my gf dumps me through docs i like lose my shit and flip my laptop forward.

User87random: That doesn't seem smart.

yaboi_Alex01: yea no shit sherlock. so im already pretty close to the front of the board. so i flip the laptop and almost hits the teacher. she got scared like real scared and sent me to the principal's office. the principal suspended me for three days. like me being suspended is going to help what.

User87random: I've always been opposed to suspending students for actions like that. What is suspending a student going to do?

yaboi_Alex01: righhhht. so my mom freaks out that i was
  suspended. and dont forget my laptop is broke.
User87random: I mean she is your mom. She loves you and
  anything that stops you from growing she sees as a threat.
yaboi_Alex01: yea right. that high school is a daycare. im at
  home pissed right, as i should be. i got dumped, failed a test,
  and suspended on the same day. all before lunch, so i was
  hella hungry and the house was like totally empty so ive
  been eating pizza bites all day.
User87random: Does not sound like a good day.
yaboi_Alex01: my mom comes back and then goes off. but the
  worst part was that she said none of my other siblings
  would ever do something like this. so there i was being
  belittled and told how much of disappointment i was. and
  now here i am.
yaboi_Alex01: like i know you're mad, but don't compare me to
  my siblings. u brought me in this world and now im trying
  to live my best. walking on eggshells for ur benefit.
User87random: I'm no parent, but I know that's not what she
  meant. Your mom was just overreacting. I'm not on her
  side, but you should know that parents speak with emotion
  and sometimes lose their head just like their children.
  People grow and learn every day until they die, never forget
  that.
yaboi_Alex01: woahhh chill with that. u sound like u have a
  secret child or something you're hiding from ur family. how
  does a college kid raise an actual kid?
User87random: Wouldn't know. Laugh out loud. Need to
  graduate first and then settle down.
User87random: I'm interested in this eggshells analogy. Care to
  explain?
yaboi_Alex01: sure ig. it's an expression i heard once, and i use it
  a lot now. basically everyone wants me to do one thing and
  act this way and only talk about these topics. im only 16 and
  im walking on eggshells and barely allowed to be me.
User87random: Stop saying your age! Do you mind if I give you
  some advice?
yaboi_Alex01: can't be worse then how im currently doing.
User87random: Step on all those eggshells.
yaboi_Alex01: wdym?
User87random: What does that mean?
yaboi_Alex01: What do you mean?

User87random: What does wdym mean?

yaboi_Alex01: What do you mean?

User87random: I literally just asked a question.

User87random: Oh, I see now.

yaboi_Alex01: lmao bro. u cant be serious. Like lmao bruh come on.

yaboi_Alex01: are you sure you're in college.

User87random: Haha very funny. I was overthinking for a moment. Now back to eggshells.

User87random: Step on the eggshells. Skate on thin ice. Do you. Live your life.

yaboi_Alex01: kinda hard when u barely have a permit and serving a suspension.

User87random: Appreciate this. This moment right now.

yaboi_Alex01: i can't wait to be out of college and have my own family, maybe.

yaboi_Alex01: what is there to appreciate?

User87random: So you got suspended. Don't let it happen again. Appreciate the fact that you cared enough about someone to actually lose your head and let your emotions take control of your mind. Appreciate the fact that this happened, not the punishment.

User87random: You will never get your time back so spend it carefully. Don't spend time lamenting over mistakes. Accept what happened and live your life.

yaboi_Alex01: good advice, but how does that relate to eggshells? :/

User87random: My bad, I got carried away. Eggshells, break them. Be respectful of course, but don't stop from being you.

yaboi_Alex01: i'll be me when i have my own family and become an insurance agent that makes a solid 6 figures.

User87random: No, you should be you now. Time won't be given back. A receding hairline will be your best example. Just tell people you don't want to live your life like that anymore. If you're doing something for your parents' sake, stop. If you're doing something for anybody else's sake, stop it.

User87random: Speak your truth. If it upsets people, then at least you know how they feel instead of avoiding topics, so they don't feel uncomfortable. Life is meant to get people out of their comfort zones.

yaboi_Alex01: yea but what if it offends communities?

User87random: ??

yaboi_Alex01: u know like the gays, queers, deers, and everything between.

User87random: That took a turn. Well now that's just plain ignorant for dismissing someone for the life they choose. If you're offending communities on purpose, then that's awful and you should immediately rectify that. I'm trying to voice concern that your life may not agree with some people because you live it the way they believe is correct. You have to live your life for what you perceive to be true.

User87random: My message: just because your friend is a Muslim doesn't mean you as a Christian can't be friends.

yaboi_Alex01: you're good bro, i can't even vote yet.

User87random: The point being is that just because you may not see eye to eye with someone on a topic doesn't mean you can't be friends or at least respect them. If that other person refuses to acknowledge the other side and simply implies that they're right, inform them that they're a bigot and ignorant instead.

yaboi_Alex01: eggshells, break them.

User87random: Live your life and speak your truth.

yaboi_Alex01: :), aye aye captain.

User87random: People believe life is going to get better if they wish it. If you're not the change you want in the world, then don't expect change to happen.

User87random: Insurance agent?

yaboi_Alex01: yea it's a stable job and u can get a salary increase just by time spent for the company in the same place.

User87random: But do you even like filing or looking at claims?

yaboi_Alex01: i won't care as long as im making bank and can afford my own man cave. i mean the cave would be nice, but as long as i can afford to spoil my 3 or 5 kids then that'll be okay as well.

User87random: Noble. 3 or 5 kids? Why not 4?

yaboi_Alex01: im not sure if i want a basketball team or an entire outfield. i can't name any activity that is good for four people besides bobsled. i dont want my kids to ever feel left out or disconnected from each other. if my kids do bobsled then i know i failed as a parent.

User87random: It could be worse.
yaboi_Alexo1: it could be a lot worse, like what if they ended up
     just like me. doing what people want me to do and saying
     certain things.
User87random: Then break some eggshells.
yaboi_Alexo1: breaking all the eggshells.
yaboi_Alexo1: u know for some old guy u really helped out
     today.
User87random: You mean grad student?
yaboi_Alexo1: tomato potato. but thanks. ;)
yaboi_Alexo1: i would stay longer, but dinner's ready and i think
     im going to break my mother's eggshells while i end my
     starvation.
User87random: Good luck.

yaboi_Alexo1 has left the chat
User87random has left the chat

The door creaked open as a blonde woman with an infant in her arms began yelling down the stairs, "Really Alex, this is what you're doing with your free time? A forty-two-year-old insurance agent, downstairs away from his family. The same family he begged to have. Let's make our own team. A basketball team he says. It'll be fun he says. Do you know what five pregnancies does to one body? Did you ever ask how I felt after?"

Her tone shifted to something sweet and gentle "Hey hon, how about you do me a favor?" She paused and then her tone went back to the one that began the conversation, "Get out of your misogyny hut and get the fuck up here. Somehow four of your kids want to see their absentee father. The other one has a wrestling practice today and he's going to be late if you don't leave soon. Don't even think about not going. Your son doesn't even like the sport. You made him do it and now he only does it to impress you. He does it for you, you asshole." The woman with a child began to shed a tear. She noticed the tear and wiped it away with the infant's dirty shirt. She was questioning how she ended up being a housewife. She had a bright future, but for some reason she threw it away. She's questioning how she ended up here.

The door slammed behind her as she left. The infant in her arms began to cry. Alex logged off and backed away from the computer in the basement of the three-story house. He turned off the lights and began his slow ascension up the stairs. His face was stuck with discontent. On the way up he contemplated about the boy online with the same name who wished for his life. The boy wanted this life, his life. Alex wondered if the boy would end up like him.

His time of thinking was coming to an end as he was preparing to face his children and disappointed wife. The discontent left Alex's face as he turned the knob and changed his face to one with a smile. The only thing heard in the house when the door was fully open were small sets of feet running towards Alex's location and the joint scream of "Daddyyyy."

LESSON 5:

The stars aren't that far or wrong

*There was a time when I topped tipped toes to touch the
mirror.
I couldn't see myself. Things haven't changed.
People look through me, not at me.*

Trisha's thoughts on the seasons

By: *Trisha Dube and Kendall*

In the summer I gave you jasmines, but you wanted roses.
So I colored the petals with my blood and etched a crimson
     pattern.
It looked a lot like roses
but then you changed your mind and asked for daisies.
So I stood there with my wilted jasmines that looked like roses
and searched for daisies instead.
You wanted lilacs, so I waited for spring to come.
Spring came, but you didn't come for me
When winter waltzed in you wondered whether the weather
     would be good enough
where you can wear what you wanted to,
and I volunteered to be your windbreaker.
Fall feebly fell through following the failures of friendships
which led to my door,
but you stayed until your faint feelings felt better
because you used me to make yourself feel fulfilled
and now there's a hole where you used to stay.
I can't read minds,
but I know you'll never tell me how you feel
now I question if I ever was real.
Real enough to acknowledge my presence instead of my
     unwanted presents
and I presently offer you my all, but it is never enough.
See I waited for you to turn your actions around,
but all you seemed to care about was some flowers.
I guess that's cute you always loved flowers,
but these flowers aren't special.
These flowers you wouldn't even cherish,
these flowers that erase the memory of me.
I picked flowers instead of my hair hoping you would accept me
Hoping you would choose me instead of this world
To think this all started with your wanting some roses
From roses I realized I'm more than your flower boy,
I've risen, arose, time got stuck so it's frozen.

I feel bad for you,
the person who wanted seasons to change and not anyone else.
If I flip the script and tell you what I want, would you get it?
Because I want geraniums and they only appear in summer,
so when summer comes will you finally sprint my way?
Will you tell me that my skin is the reason why you loved
     flowers instead of me?
I'm glad that I was almost lost to you,
it made me realize that to myself I must remain true.

Modern day abecedarian with eight people

At eight hours *before* eight a.m. on a Sunday there were seven
  boys and one girl outside,
But boys believe that boners *can* be placed anywhere with no
  consent given.
Conscious or unconscious consent was not given a *damn,*
"Dicks deserve to be placed *effortlessly, even* if she was drunk
  and barely awake," they said,
Eternal damnation for anyone who *finds flaws* with what they
  did.
Faithful boys did a foul deed indeed, but *God* loves a
  remorseful sinner.
Gleefully the boys were glad to see the priest who says *hell* is
  not where they would go,
Hell was a term meant to scare atheists, not good boys *inside*
  churches.
Instead of the boys being charged, the girl was made to leave
  her home in late *July.*
Justice would not be served, *kind* of like three meals on
  Sundays.
Kindness was not shown by friends and family to a girl who *lost*
  her dignity and *love* of *life,*
Last thing she remembered was pain and the only other person
  blamed was her *mother.*
Mothers *need* to teach daughters *not* to trust boys when
  drinking, she's to blame.
Never mind that the neighbors saw the boys doing this evil act
  outside and they said nothing,
Of course the blame is still on the *porcelain* mother who never
  thought this could happen.
Pity the mother who saw everyone *quickly* turn their back on
  her and her daughter.
Questions answered and boys told lies, but *regret* and *remorse*
  were not in their tones;
Revenge would not be taken, only thing that was taken was a
  girl's life due to *suicide.*
She screamed from sadness as she was not sober but was
  tormented.

Turns were taken as she was tormented that night and she
    screamed every night *until* her life was gone.
YouTube accounts were made to *vlog* about the incident,
Videos *were* shared *within* communities and anyone there that
    night;
Wait, but still the girl was blamed, and her body was *examined*
    and it showed that she resisted.
External crisis led a girl to suicide, this is now a *yearly*
    occurrence for some people.
Yet here we are back at *zero* because the mother also yelled in
    agony as she took her own life.
Zero, zilch, nada, and nothing happened to the boys for what
    they did *eight* hours before *eight a.m.*

Confrontation

I don't like confrontation
My grandfather was a police officer, I never met him
He decided to not confront his heart when it needed
    confronting
I don't like confrontation
So when she touched my hair and played with it,
I just felt embarrassed,
When she started telling other people to feel my hair,
It felt like something that was a part of me didn't belong to me
I just don't like confrontation and if I did,
I would have to explain it's impolite and rude to touch a black
    person's hair;
fuck that, anybody's hair.
At work another woman told me how glad she was that her son
    wasn't black
Her intentions were good, but I question why she told me
As if telling a black person absolves you of your non-existent
    guilt
for not aiding our fight to end systemic oppression
instead, you said silent prayers under your breath when you
    went to sleep
knowing our issues could never happen to you and yours
Our size 15 clown shoes that have shackles around our ankles
and chains on our neck that are decorated with yellow paint to
    resemble gold
We wonder whether the whips artists talk about in songs
are that much different from those that used to terrorize the
    backs
of the ones who walked around barefoot and chained
I don't like confrontation as I tell this to my non-black friends
    who want to say the *n* word,
the nigger word
When they attempt to say the word, I leave.
I do not wish to confront them on their habits
and start a conversation in which they will never relent their
    side
But growing up colored I was taught giving up might be the
    only way to advance.

That's what we as a community do;
when police officers shoot unarmed people and justice is not
    served,
we accept defeat so that we can cherish the moments we are
    allowed to live
We believe the lie we tell ourselves so that we can try to
    breathe,
like everyone else
Prayers at family dinners consist of everyone saying the names
    for the people at the table,
praying that we see them tomorrow,
alive.
I don't like confrontation, as the lady who knows me asks:
what it was like to apply to college knowing that I had a better
    chance than the "other kids"?
Gleefully, I tell her that I think the system is broken
I wanted to tell her that the "other kids" won't understand
Were they admitted off merit or did the school need diversity?
Some are fine with not knowing,
but when you are told this is the one time being colored helps,
what are you to think?
I don't have to say the color of the skin, or the length of their
    hair,
or even the shape of eyes, or describe the lips
For each person, we all know who this is and what they look
    like
I took a look back in my life and realized that I love
    confrontation
I just don't like confronting people who see oppression as a past
    action
and not a modern-day current event

A.D.H.D.

Hair it was heavenly
Beautiful was her body
Eyes were elliptical
Her mouth was mesmerizing
She was a dime a dozen in a pack of nickels
Could crack cards with crooked cops and crooks
called them Cruikshanks
Justly they jaggedly jawed and jostled while
jarring and jeering as she glistened
Remarks, remade, rewoven, rewritten, and returned
as she slipped a slippery smile
She showed a smile that had saliva still sautéed in the smell of
    single squares
Notice how nicotine never knocked her down,
it narrowly narrowed her nonsense
Newport's never nudged her in the right direction
Caught with Camels not caramel,
wish it were chocolate like her skin but guess not
She always looked west when the sun rose
when it fell, she always claimed that she missed it rising
And would end up talking to toothless
tone death
two faced toddlers
two decades past ten
She resented her mother for her imperfections,
always perfecting pinpoint procedures of beauty
Needless to say, no one knew she didn't know she was beautiful
They couldn't tell,
more like they can't read when what she did was spell
She spelled out what was wrong and
to her demise and downfall they just gave her a song
A song that say she's beautiful no matter what they say
I'm sorry Ms. Alisha
but this girl didn't have the keys to get out of her own head
This girl was on fire, but not in a good way
Besides that, her energy was often exuberant
but euthanized before she could even exercise
Inattention to accurately accrue information

as she dozed in between decimals and demons
However, it doesn't matter how it happened because she hid
    how she felt
How her emotions were entangled and mind often mangled
It's hard to walk straight
when your eyes look every direction but forward
This girl was amazing when she was all and attentive
No hocus pocus, or magic,
but she went into a box and never came out
She had A.D.H.D.
I prefer to call it as it was,
the proper abbreviation that should be used for her:
Addicted, definitely hiding demons

The Universe Gives Great Advice

Contrary to popular belief I am not a hopeless romantic. In fact, I am just a helpless idiot. I believe with my whole heart that I should come with a label that reads "enter at your own risk" or "caution."

I've decided to break the stereotype and tell the truth of what happens when I leave a "relationship." I wonder if you can even call it that. Most girls who have been with me try to bury that part of their life away like treasure. The only time I am ever mentioned is when someone asks stories about who is your worst ex. I was always taught to be the smartest man in the room, no one ever said be the smartest person in a room. Somehow, I ended up dating girls who are always smarter than me and generally the smartest person in the room or any scenario.

My mind is a maze and all my previous "girlfriends" tried to navigate it. I warned them not to. However, they didn't listen and found out that the maze was littered with landmines and ticking time bombs.

There are three things to keep in mind: I am charming as hell without knowing it, none of my exes are crazy or close to it (anymore), and I always thought that I was mad at them for leaving without any sane rationale, but they left because I never loved myself enough.

Instead of taking accountability for my actions, I would like to specifically blame my mother for showing me the Hallmark channel.

Hallmark movies with my mother never quite helped. I would do my homework on the side of her bed and join her as she walloped when the main character missed the cue that the handsome, yet average looking white man was hitting on her. I learned a lot from those Hallmark movies. I used my knowledge from Hallmark movies on my "girlfriends". I don't even like that term, girlfriend. I prefer my girl, my lady, or anything that signified she was with me and didn't belong to me. To be precise, there were two girls who made me realize that I am a helpless idiot.

The first of two girls to be mentioned is Ariel.

Ariel was as close to a Disney princess that I would ever date. I would like to say it was magical, but it was horrendous. By horrendous, I mean for her. By magical, I mean for me. She made herself available any time I asked with no reservation. She made sure I went to everything on my schedule. She hated my dirty laundry, so she did it for me. When it came to interviews, she picked out my outfits, and she never complained. Correction, she never complained to me. We only lasted for a month and a half but somehow, we went through everything that a failed marriage experienced.

I call her a Disney princess, because she was like a princess in every Disney movie. She loved animals and animals loved her. Ariel wanted to make the world a better place, and was always looking out for people who didn't deserve her compassion (me). From the minute we got together, all her friends told her she was going to regret the decision. Ariel didn't listen. What happened was that I did exactly what her friends warned her I would do. I didn't try to sabotage this relationship. Things just happened, and I let them happen.

To be honest, I don't know why she stayed with me. For example: I joked that I had a kid. I told her that if she got pregnant, I would disappear and never be heard from again. And we never did anything in public. Sometimes I asked her if she thought another girl who flirted with me was cute. Looking back to Ariel, I can't imagine why she stayed.

She was cute, had freckles and was a brunette with a tinge of blonde. The height difference between us was perfect. Her dream was to travel the world and give smiles to those who can't afford basic necessities. I briefly died when she told me her plans. She was incredibly cool and just loved life. Awesome person to know, highly recommend: 12/10.

We only made it a month and a half. How we made it that long, I don't know. Her friends would always complain to me that she was crying because of something I said or did. Ariel never told me. I denied her friends' claims and that sounds awful. How would you react if people who didn't like you and never wanted you to get together told you this? Would you trust their words?

Ariel did everything right, and I did everything wrong. She was concerned that I would never truly care for her because I didn't care enough about myself. The idea sounded ridiculous to me as I told her that I would take a bullet for anyone. That didn't make things any better. We were standing outside of her place when our final conversation happened. It was her birthday.

"I would take a bullet for you. For your roommate who hates

me. For anyone in your family who hates the idea of you being with someone who doesn't look like any of your past boyfriends. You seem mad. What did I do?"

"There it is again. You do that dumb shit again. Your trademark dumb shit." Her voice cracked and became shaky. "You can't offer up your life every instance. Why am I mad? Maybe it's wrong for me to think selfishly, but you have to be selfish. Just a little for me. Before you want to jump in front of a bullet, don't you want to think about all the people who you're going to leave behind? Your siblings, your mom, and me. With me you have to be a little selfish." Ariel paused to gather her thoughts and let the lump in her throat die down. "I'm sorry, but that's me. I love the fact that you want to be this selfless guy and try and save the day. But if there's no hesitation before you save someone's life then I can't be there for you."

"I didn't ask you to be here. You chose to be and continue to stay. I'm not making you do anything."

"You're making me care about you. When you're not in my bed I question if you're going to wake up breathing. I wonder if you're going to overthink everything and try to end your life. If that happened how would I explain that to your mother. The same mother who you never talk about and the same one I've never met. My entire family knows about your existence and how special you are to me." Her eyes were red and gushing with water.

The freckles disappeared. She wasn't blushing, she was flustered. Her hands moved from her side and shifted as they transformed into fists as she began to pound my chest and continue crying. Ariel was not strong, it felt like someone was trying to play the drums on my chest. She stopped, inhaled a deep breath, and wiped her face with the sweater I was wearing. She left her fists on my chest and looked up at me. "You can't do that to me. You can't tell me how you would do all these things for other people but not put in an effort for me."

"I do put in effort."

"No, you just give explanations for why you do the things you do and hope that I accept them. Well I don't. I don't agree that you have to help everyone you see. You're not everyone's hero. You're you. I'm sorry that's not good enough for you to accept. You're worth a damn and the minute you realize that you'll understand what I see in you. You'll understand why I stay with you even when you make me feel like shit."

"The only hero I got was my mom." I moved my hands to try and grab her fists which were still on my chest. She moved quicker than I did and folded her arms. "If you could help as many people as possible and they don't even know your name, wouldn't you do that?"

Ariel took two steps backwards and hit the wall. I took one step forward and wrapped my arms around her small body. She didn't try to escape my embrace; she accepted it and nestled her small head underneath my chin. That lasted for a small moment before she rejected my embrace.

"No, I wouldn't, and I guess that's where we stand. I can't keep doing this to myself. I just can't do this to myself anymore. I'm going to do something that I think you should try, and that's looking out for me as a priority. I think we should end it." In between the gasps of air to calm herself down she made her final statement, "It looks good on you, but I want it back."

Ariel was referring to the necklace I was wearing. To be honest, I forgot that it was around my neck. It was the first thing I stole from her when I spent the night at her place. She was right. The necklace looked really good on me. It exaggerated my Adam's apple.

Ariel and I never really spoke again. We hooked up once after that day and talked about trying again. Nothing came from it. After me she deserves as much happiness as she's allowed. I never realized that caring too much about other people would be my downfall.

After the unsuccessful "relationship" with Ariel, I thought it was the universe telling me something. I think the universe was trying to tell me that maybe I should focus on making myself worthy of my own self-love. If the universe was telling me that, then it did a bad job and the message was not received.

The second girl to be mentioned is Briana.

Briana was different. Unlike her natural demeanor, she was as fierce as a volcano ready to erupt. She was from one of the U.S. territories, one of those islands. She told me, but I couldn't remember. She told me the full names of all of her family, friends and I got lost in confusion and translation. She told me she was mixed with a lot of different races. She was part black, a sixteenth native, 21.2% Scandinavian and the rest I couldn't name if a gun were put to my temple.

Somehow she spoke Spanish, her tiny body hairs on her arms and legs were blonde, and she was an amazing kisser. Her hair was curly brown and bounced but never lost its shape. She loved wearing my shoes around and being clumsy; she always questioned what it was like to walk around with big feet every day. I'm not tall, but she made me feel big.

I joked with a friend that I was dating myself. Like myself, Briana was unapologetically herself even when she should've apologized. We always fought and there was never an apology. People questioned if we were actually happy together. Most days we were.

The reason why we fought? When you date someone that reminds you of you, you start to realize all the parts you don't like about yourself. Dating yourself shows you all your flaws along with the hidden beauty. Dating yourself also becomes a fascination. You become fascinated if this is how you appear to the world. I don't recommend dating yourself unless you're ready to see what and who you really are in a relationship. With that being said, I would do it all over again.

Briana and I lasted for about three months. It was a summer fling. The single reason we lasted so long, and I allowed myself to be dumped was one of the first lines she ever said to me. I don't know what we were talking about or how this came up. To this day I don't know if it's a compliment or insult. She told me, "You're a helluva a man." I don't even remember the rest of my day. I don't know if that was said in the morning or the night.

Briana was 5'3" and was absolutely beautiful. My slight speech impediment never allowed me to say her full name correctly. Somewhere I would mess up and then she would laugh. The laugh wasn't at me, it was at my attempt (I told myself). To avoid this problem, I always looked directly at her when I wanted her attention and spoke.

If you raised your voice at Briana, she would raise hers ten times higher. She was different than any other girl I was ever with. I noticed some key differences between this island girl and anyone else. For starters, she would call you out on anything she didn't think fits the current scenario. She constantly thought I was losing interest in her when I was just tired and really bad at communication.

The other big difference I encountered was what happened when Briana was mad. Unlike normal girls who may ball their fists and at extremes lightly jab you, she unleashed her rage. We once got in a fight on whether or not cereal was a type of soup. She got so mad that she picked up my gaming console and threatened to throw it out the window unless I admitted she was right. We stared at each for ten minutes in silence before we went out to get Korean food. I feel that it would be pertinent to add that she threw my toaster at the wall one time. Why? Because I told her I didn't want to meet her parents. We were only three weeks into the relationship; it felt very early.

Briana gave me my first bloody nose and also had to take me to the hospital because I wouldn't stop bleeding. I had to get stitches. The nosebleed wasn't her fault, I tripped on the stairs and fell. How did I trip? I wasn't paying attention and she scared the crap out of me. Briana was there with me the entire process and sat with a furrowed brow and arms folded while I was being operated on. I never accused or blamed her for anything. In return she said that I should be lucky she didn't ask me to clean up the blood that dripped on her passenger seat. Briana drove a Range Rover; it was part of her personality.

Most people would question why I didn't leave after those moments. When you find someone who can match your energy and craziness, you don't really think about leaving. She had so much passion for something. I prefer people who actually give a damn rather than the ones who just float along in life.

We argued a lot, but that was just our thing. We were both so passionate about life and never wanted to relent our way of thinking. I didn't see our final conversation happening. I didn't see it coming at a Chick-Fil-A. The A in Chick-Fil-A should stand for awestruck because what the actual fuck.

The final conversation started with me complimenting her eyes and being as cheesy as a burger.

"Your eyes remind me of the stars."

"I know. My daddy was an astrologist and put the stars in my eyes."

"No wonder I see constellations when I consider our future."

"What are you doing?" she said. Her cheeks lost the ability to hold blood. They went brown like the rest of her skin.

"I thought I was being cute." I flashed a quick wink and attempted to grab her hands. She pulled her hands back and placed them on her lap before I could even touch her outstretched thumb.

"Stop being you for like one minute and listen to me. You

tried to save that little boy's ball."

"So that's what this is about. What happened yesterday? The beach was fun."

"You tried to save that little boy's ball, and you can't even swim. You didn't come up for like two minutes. Do you know what I went through or what I was thinking? I was at a loss of words and couldn't breathe. Normally when you do dumb shit, I'm the only one who worries. This time you had that entire section of the beach worrying if you just drowned trying to save a ball. "

"Well I got the ball. I already told you, the current was a little strong and it took me a while to get to the surface."

"That's not the point. You can't swim. Why not let the ball go? A ball is not worth your life. That family could afford extras."

"I mean, I figured when it counts, I would learn to swim in a life or death moment."

"You can't just go off and try to be Superman. You were right next to me on the towels. I turned around for ten seconds to hand you a bottle of water from the bag. And guess what, you're fucking gone. Up to your neck in water and only going further. I wanted to rush after you, but I couldn't move. My heart dropped and I was stuck. I was paralyzed with fear. Fear for and from you." Her hands left her lap and slammed the clean white table. People turned their heads to see the commotion. "If you died then how was I supposed to get back from the beach? I didn't know where you put your keys. Ubers don't go that far. I am only joking right now, because I don't think you understand the seriousness of that moment. You didn't think about anyone else and just acted off instinct."

"The kid seemed really happy and the mom tried to pay me." I blushed talking about my accomplishments.

"Do you feel happy? Do you feel good?" She was visibly upset. Her voice wasn't as stoic as before.

"With you, I feel all that and invincible."

"Good, because I want you to know that I am broken. You broke me. And I never want to be broken again." On cue, she started emitting saltwater from her eyes. It was a hushed cry. She didn't want any more prying eyes, so she spoke in a low voice, "I saw it all with you. The bullshit family with a picket fence, but no dog because you don't like dogs. Rotating Christmas and Thanksgiving dinners. Having you sit with my family at my graduation. The fights over what middle name we give our kids.

The surprise weekend trips. You asking my family for their blessing to start our life together. But you don't care." I grabbed a napkin and tried wiping her tears away. She snatched the napkin out of my hand and did the job herself.

"If I broke you, then let me mend you back together. I never said I didn't want any of that other stuff with you." My face remained calm and my voice followed suit. "I can see all that with you, just let me show you."

"You don't get it," Briana hissed. "That's not some other stuff. That's *the* life. The life I want to live. I don't know what to do with you. I can't sit back anymore and watch the boy I have deep feelings for self-implode. There's going to be one day where you can't save the day and you're just going to end up dead. And I can't be there when that happens."

"I can promise that as long as you're by my side I will never end up dead." I attempted to grab her hands that were on the table, but once again she moved them. She grabbed another napkin.

"But you can't promise that. As much as you think you can command death, you can't. You don't know when death is going to happen. When we were driving to that Korean restaurant and we saw an accident. What did you do?" A sly grin came across my face. "See that's my point, it's not funny. It was serious. The accident was on the other side of the road. You pulled over—"

"Safely," I chimed in.

"Yes, safely. You pulled over, leapt out the door, and sprinted across all lanes of traffic. That sounds so heroic unless you were the person who was left in the car watching. You didn't even check if cars were coming. You just ran with no thought and no care in the world. And guess what, the man was okay and had already called an ambulance. You weren't needed. You could've been hurt or killed. But you didn't care then, and you don't care now. Why can't you just give a damn about your life? Put yourself first and realize that someone has you in their future. Someone has you in their dreams."

"I'm just doing what I would want done."

"And that's great, but not everyone is like you. The world might be better if it was, or scarier, but that's how life is."

"Then why can't I change life?"

"You can't change life."

"So what now? I don't get why you don't want me to be the change I want to see in the world."

"I think we should call it. See, you're trying to change the world, but never yourself. Growth, you don't want that. You clearly want to save the world rather than yourself, and I can't be there. Correction, won't be there."

"That's it. Nothing else."

"Unless you're going to stop being you. I fell for that guy. I didn't agree to be at your side and watch the many ways you try to kill yourself."

"But like we ordered food and like I paid for it."

"Cabron, no me importa. Vete." I didn't move. I stayed in my seat. "Do I have to say it in English? Leave. Go. We're done." I scooted out the booth, I tried not to look at the people who were looking at me leave. As I left the booth, the food I paid for arrived.

Since that day in Chick-Fil-A, I haven't spoken to Briana. I noticed her Instagram; she appears to be happily single. During our "relationship" we talked about taking a trip somewhere sunny with a lot of beaches. I still get the notifications for when prices are low, and I get tempted to buy seats for two every time. I want to say she held a special place in my heart.

Ariel and Briana come from two opposite ends of the spectrum in terms of personality. However, they both crossed my path because the universe had our fates intertwined and interwoven. In the end, I think the universe was sending me another message with bright neon lights. If the universe was in fact telling me to make myself a priority and actually care about what happens to myself, then my best response would be: maybe. Maybe I'm actually blind.

BONUS POEMS

Dreams

My only nightmare is leaving,
but I guess it's reality cuz I already left
Falling from gravity whoops goes reality and fell into actuality
a place where your loved paid off salary but this ain't
accidentally
and whoa I see Billy
Billy's the boy who wears bombers
and breathes boxed wine to block backdoors
and break down barriers barricading bars
but whoops goes reality as he falls into gravity.
And sadly we have no other reason for posterity and this is a
rarity. . .
A nightmare isn't supposed to last a lifetime,
yet with you I see no clarity
I may be dreaming but at least in this world I'm not a villain
Villanelles as bland as vanilla see you demonized me
Made me question if the stars were still shining for free
Made me wonder if my shadow deserved to follow me
But who are you to make me feel this way?
You are anyone who made me question not only reality
but faith
Faith that the world only produces evil monsters
I prefer believing that no matter what I do it will never be good
enough
Not good enough for me
and definitely not good enough for you
But in my dreams, you don't matter
Not now,
not next week,
or even next month
But someday,
you won't matter
When that happens
my dreams won't just be dreams anymore
But for now I'll finish my dream. . .

Whoops there goes reality
and now I'm unstuck from an abnormality
where my personality isn't considered insanity or brutality
But hey, pinch me, I think I'm dreaming

More Flowers

Lavender lilacs lay less with lethargy lamenting over loss.
Sweet sunflower symphonies turned sour, sinful, and sinister.
Delicate daisies dawned dark dreams.
Gardenias gloat glees of glamorous gluttony.
Repugnant roses rip and rot relationships.
Vicious violets vehemently violated the vicinity vertically with
vertigo.
Marigolds may very well make mayhem mimic mischief
manically.
Carnations copy capricious cycles of cynical chaos.

When I look at flowers, they don't seem beautiful anymore.
They remind me of how much you loved flowers.
They remind me of how you wanted to open a shop,
They remind me of you dying,
where you weren't just fighting, you were trying.
I hate flowers, those in particular.
I hate that they're the only thing that reminds me of you.
I hate myself for not loving the flowers,
the ones you loved.
I want to love myself and the flowers
like I loved you:
With all my heart

Elephants in the room

Wherefore art thou, "friend"
Shakespeare couldn't wish for a better actor
to act like they care
Hiding behind witty wishful words and
phony prophecies like fake prophets
Big and ugly
Small and pretty
It's all the same
Try to find the words to say this world isn't shitty
It's tough to admit reality,
We often hope that we are dreaming because
this life doesn't consider equality
Doesn't it suck,
when there's a double standard not only off gender but also off
    skin
Doesn't it hurt,
when resisting is only considered in darker shades
Doesn't it make you sick,
when life throws you a bone but it's smaller than everyone
    else's
Don't leave, don't leave me
Don't leave my side when wrong is being done and this isn't
    right,
Right in front of my eyes
Correction, in front of our eyes
Ask me what's wrong
Ask me what I want to change
Don't ask me what the day was like
and not mean it
You keep leaving after I tell you
You are my "friends,"
neighbors,
and guides
but you disappear faster than Houdini
Magic was always in your bones,
You made my misery vanish into thin air when you promised
    change,
But it's still a dream

and the destination is far, far away
No fairy godmother can make this better
For if you wish and not work to make a problem go away
Then you are leaving me with not one,
But many elephants in the room

Thank you

Eight letters,
two syllables
Sometimes two words and
other times one giant word,
The sun shined smiled this morning like it does every morning
The moon watches our actions and doesn't interfere
The stars still dance every night whether we see them or not
No one knows if you smiled this morning or danced at night,
Not everyone is fortunate with the capabilities
Not everyone can stare at the sun until their eyes begin to black
Not everyone can utter sounds of jubilation or disappointment
Not everyone can stand on their own and dance freely
Today may not be your day, your week, month, or maybe your year
And no one will go through that pain more than the person
     who doesn't want to be there
People promise better days,
but not every person is a psychic
No one knows what tomorrow holds
And this may sound dull and bleak when life isn't going your way, but
be thankful
Climbing mountains often have moments when everything
     goes wrong and seems impossible
You may not make the most,
look the best, or
be in the best shape but at least you know that
At least you wake up,
Every morning say thank you,
it doesn't have to be at something or someone
just speak it into existence
That somehow you're thankful for the life you have and
thankful for the life you want to make
No matter what happens, the sun will still smile,
the moon will still stare and watch, and
the stars will still dance
Thank you

About Atmosphere Press

Atmosphere Press is an independent, full-service publisher for excellent books in all genres and for all audiences. Learn more about what we do at atmospherepress.com.

We encourage you to check out some of Atmosphere's latest releases, which are available at Amazon.com and via order from your local bookstore:

A Synonym for Home, poetry by Kimberly Jarchow
The Cry of Being Born, poetry by Carol Mariano
This Side of Babylon, a novel by James Stoia
Within the Gray, a novel by Jenna Ashlyn
Where No Man Pursueth, a novel by Micheal E. Jimerson
Here's Waldo, a novel by Nick Olson
Tales of Little Egypt, a historical novel by James Gilbert
An Ambiguous Grief, a memoir by Dominique Hunter
For a Better Life, a novel by Julia Reid Galosy
Big Man Small Europe, poetry by Tristan Niskanen
The Hidden Life, a novel by Robert Castle
Big Beasts, a novel by Patrick Scott
Alvarado, a novel by John W. Horton III
Nothing to Get Nostalgic About, a novel by Eddie Brophy
GROW: A Jack and Lake Creek Book, a novel by Chris S McGee
Home is Not This Body, a novel by Karahn Washington
Whose Mary Kate, a novel by Jane Leclere Doyle
Stuck and Drunk in Shadyside, a novel by M. Byerly
These Things Happen, a novel by Chris Caldwell
Blood of the True Believer, a novel by Brandann R. Hill-Mann
Geometry of Fire, nonfiction by Paul Warmbier
In the Cloakroom of Proper Musing, a lyric narrative by Kristina Moriconi
The Dark Secrets of Barth and Williams College: A Comedy in Two Semesters, a novel by Glen Weissenberger
Lucid_Malware.zip, poetry by Dylan Sonderman
The Glorious Between, a novel by Doug Reid

About the Author

Kendall Nichols was born and raised in the Chicagoland area. The sixth out of seven children who had personalities as fierce as volcanoes, Kendall is a person who lives to love the joy of life and isn't afraid to speak his mind. Typically the shortest person in the room, his personality makes him feel like a giant. If you met Kendall, you won't forget him anytime soon.

www.ingramcontent.com/pod-product-compliance
Lightning Source LLC
Chambersburg PA
CBHW031350060726
47590CB00007B/2715